AF657187

ENVIRONMENTAL SCIENCES FOR ENGINEERING UNDERGRADUATES

(As per AICTE notified syllabus)

DR. SUSHMITHA BASKAR
DR. R. BASKAR

UNICORN BOOKS

UNICORN BOOKS
E-mail: unicornbooks@vsnl.com
Website: www.unicornbooks.in • www.kidscorner.in

Distributors
Pustak Mahal, Delhi
J-3/16, Daryaganj, New Delhi-110002
☎ 23276539, 23272783, 23272784 • *Fax:* 011-23260518
E-mail: info@pustakmahal.com • *Website:* www.pustakmahal.com

London Office
5, Roddell Court, Bath Road, Slough SL3OQ J, England
E-mail: pustakmahaluk@pustakmahal.com

Sales Centre
10-B, Netaji Subhash Marg, Daryaganj, New Delhi-110002
☎ 23268292, 23268293, 23279900 • *Fax:* 011-23280567

Branch Offices
Bangalore: ☎ 22234025
E-mail: pmblr@sancharnet.in • pustak@sancharnet.in
Mumbai: ☎ 22010941
E-mail: rapidex@bom5.vsnl.net.in
Patna: ☎ 3294193 • *Telefax:* 0612-2302719
E-mail: rapidexptn@rediffmail.com
Hyderabad: *Telefax:* 040-24737290
E-mail: pustakmahalhyd@yahoo.co.in

© Copyright : Authors

ISBN 978-81-7806-125-2

Edition : 2007

All rights reserved. No part of this publications may be reproduced, stored in a retrieval system or transmitted, in any form or by any means, electronic, mechanical, photocopying, recording or otherwise, without the prior permission in writing from the author/publisher.

Printed at : Pushp Print Services

Dedicated to

Our Parents

and

Teachers

SYLLABUS FOR ENVIRONMENTAL STUDIES FOR UNDERGRADUATE COURSES OF ALL BRANCHES OF ENGINEERING

(As per AICTE notified syllabus)

Environment and Ecology

➢ General

Introduction, components of the environment, environmental degradation

➢ Ecology

Components of ecology, ecological balance and consequences of change, principles of environmental impact assessment

➢ Air pollution and control

Atmospheric composition, energy balance, climate, weather, dispersion, sources and effects of pollutants, primary and secondary pollutants, greenhouse effect, depletion of ozone layer, standards and control measures

➢ Water pollution and control

Hydrosphere, natural water, pollutants, their origin and effects, river/lake/ground water pollution, standards and control

➢ Land pollution

Lithosphere, pollutants (municipal, industrial, commercial, agricultural, hazardous solid wastes), their origin and effects, collection and disposal of solid wastes, recovery and conversion methods

➢ Noise pollution

Sources, effects, standards and control

Contents

Preface

Developing an understanding of environmental issues is the first step towards environmental awareness. We are in a mad frenzy of creating a technological wonder world to live a more comfortable life but in the process the global economic activities have adversely affected our environment beyond our imagination. Urbanization, over exploitation of ground water, declining biodiversity, unprecedented population explosion, improper waste disposal mechanisms, pollution of air, water, soil etc are reaching threatening proportions. It is high time we realize the importance of preserving our environment. We fail to understand that nature is not just for us but also for all the species on the earth. It is rightly observed by Mahatma Gandhi that, *'There is enough on this planet for everybody's needs but not anybody's greed'*. It is unfortunate that in the name of development, we are damaging our wildlife and biodiversity. We should use our resources sustainably and conserve it for our future generations. Proper management of the environment is an important step towards attaining sustainable development. We must try to learn about our earth, its functioning, how it adapts to changing conditions, sustainable use of our natural resources, respect for all life forms and how we can contribute to saving our planet. Public awareness, involvement and the political will of the government can together bring about the desired changes. Greening of the body, mind and soul is essential, if we have to safely hand over mother earth to future generations.

Following the directive of the honorable Supreme Court of India, the AICTE and UGC introduced a mandatory course in environmental studies for all the undergraduate courses. This book is specifically designed on the basis of the model curriculum prescribed by the AICTE for all engineering undergraduates in India. Great Care has been taken to incorporate case studies and interesting facts to trigger the mind of the students. This book will be very useful for all students preparing for competitive examinations and to readers interested to get an overview of the environmental issues.

This textbook is divided into six units and covers a range of topics. Each unit focuses on a specific theme. The first unit is a general unit and covers aspects regarding the importance of our environment, the components of the environment, environmental degradation and an overview of the environmental problems we are facing today. The second unit gives a basic understanding of ecology, and the principles of environmental impact assessment. Unit three introduces the students to the atmospheric composition, energy balance, climate, weather, dispersion, sources and effects of air pollutants and their control. In the fourth unit, a description of the hydrosphere, water pollution aspects and its control is given. The fifth unit elaborates on the lithosphere, pollutants that affect our land, the collection and disposal of solid wastes, the recovery and conversion methods. The various aspects regarding noise pollution, its effects and control is explained in the last unit.

One of our objectives in writing this book is to motivate the younger generations to act and think in an eco-friendly way and to sensitize them to some of the important environmental issues. We would welcome suggestions, constructive criticism of the book so that the future editions can be improved upon.

We express our deep gratitude to our parents and teachers. We thank our son, Siddharth Baskar for cooperation during the entire course of work.

We thank Unicorn books, New Delhi, for their efforts to bring out our work and present it in an impressive way.

—Dr. Sushmitha Baskar

—Dr. R. Baskar

UNIT 1

GENERAL

Introduction

Nature affects mankind and human endeavours interact with nature in every region of the globe. Nature has not suddenly stepped up its activities, but human populations have increased dramatically and the various technologies we use today require ever-increasing inputs of energy, space and resources. Before 1900, no city of five million existed. The tremendous impact of human activities has changed the physical environment and has put it out of balance with nature. Understanding our environment and the processes that control it is one part of the puzzle. Sustainable solutions should accommodate the present social, economic and cultural needs without compromising those of the future generations.

The term environment means surroundings. Our environment is the entire web of biological and physical interactions, which characterize the relationships between life and the earth. Environmental science is an interdisciplinary subject and involves the study of various subjects in relation to the environment. It helps in understanding a scientific basis for establishing standards for acceptably clean, safe, healthy environment for humans and the natural ecosystem. Environmental science starts by understanding how the natural world in which we live works.

The Environment Protection Act, approved by the Government of India in 1986, defines the environment as the *sum total of water, air, land, the interrelationships among themselves, with human beings and with other organisms and property*.

Environmental awareness among people is important and it involves developing an understanding towards environmental problems, the way we respond to our environment and adopting eco-friendly lifestyles. We are a traditionally rich country and even today we worship nature (i.e. *Tulsi puja, Surya namaskar* etc). Nevertheless, during the past few years, there has been a gradual decline in our natural resources and the quality of our environment. We have put an increasing stress on nature, with growing demands for food, water, mineral and energy due to increasing population and changing lifestyles. Environmental problems, issues and challenges are complex and multifaceted. The survival and progress of a country depends on the sustainable use of resources. Today our newspapers, televisions and FM radios highlight the environmental problems we are facing, which was not the case a couple of years ago. All this is needed, as the general public should be made aware of the importance of environmental issues. An example worth emulating is the government of Himachal that imposed a ban on plastics in 1998 and today polythene and plastic bags are not seen in the whole state. It shows the role of public cooperation with the government in conserving the environment.

Before 1960, very few people had heard about the word environment and ecology. Then, came a landmark publication Silent Spring (1962), by the famous biologist writer and ecologist Rachel Carson, which triggered off the environmental awareness movement all over the world. Carson's work showed that DDT and other chemicals used to enhance agricultural productivity were toxic and warned the general public about the long-term effects of misusing pesticides. Her work challenged the practices of the agricultural scientists and the government, and called for a change in the way we perceive nature. This was coupled with the major oil spills and threats to endangered species. Following this, the environment became an important subject and received worldwide attention and focus. The Stockholm Conference on Human Environment (Sweden, 1972), The United Nations Conference on Environment and Development (Rio de Janerio, 1992), The World Summit on Sustainable Development (Johannesburg, 2002), have made the public aware regarding the detioration of our earth.

As far as the Indian context is concerned, Article 48 A and 51 A (g) in the section on directive principles of state policies in The Indian Constitution, approved in the year 1976 states that the "State *shall endeavor to protect and improve the environment and to safeguard the forests and wildlife in the country and to protect and improve the natural environment including the forests, lakes, rivers and wildlife and to have compassion for the living creatures*." Despite this the importance of Environmental Studies was not recognized in the Indian curriculum. Hence, the Honorable Supreme Court of India in 1991, on a public litigation interest filed by Sh. M.C. Mehta in 1988, made it mandatory to include the subject environmental studies in all the universities and colleges. This was basically to create environmental awareness among the Indian citizens.

Major global environmental problems include global warming, stratospheric ozone depletion, acid precipitation, deforestation, hazardous waste, increasing human population, threat to biodiversity and depletion of our energy resources. Also, the impacts of mining, construction of dams and their socio-economic aspects, pollution and waste management are very important environmental issues. It is important for us to understand the transport and fate of pollutants and toxins in air/water/soil. Chemicals introduced to the marine environment as a result of human activities include toxic substances that generally exert their effects on organisms through direct toxicity and the extent of these effects are dependent on the chemical nature and concentration of the substance and the period of exposure. These bring about changes in the natural balance of the physico-biochemical environment through a series of physico-chemical and biological transformations.

Proper management of the environment includes the conservation and enhancement of biodiversity and maintenance of ecological processes and life support systems essential for a functional biosphere, sustainable management and development, to promote and encourage education, training, study and research in environmental science. The Chinese philosopher quote mentioned below is relevant in this context.

'If you plan for one year plant rice
If you plan for 10 years plant trees
If you plan for 100 years educate mankind'........Chinese philosopher Kwandu zu

The problems faced today are directly or indirectly related to the environment and solutions need a sound environmental knowledge. Human-induced environmental change has accelerated over the last three decades because of advanced technologies. The increasing complexity of the environmental degradation requires an enhanced capacity for scientific awareness, assessment, monitoring and early warning. It is necessary to sensitize environmental issues right from school children to adults, as our approach to environmental issues is the way we treat our environment and is basically an attitude towards way of living.

Today the issues such as global warming, climate change, pollution control and natural resource development and management increasingly require the crossing of established boundaries between disciplines. Technological solutions are also needed to minimize the impacts of environmental pollution. Environmental Engineering involves the study of the technical processes that are used to minimise pollution.

Components of the environment

The various components of the environment consist of the atmosphere, lithosphere, biosphere and the hydrosphere. The atmosphere, solar energy, and our planet's magnetic fields support life on Earth. The atmosphere absorbs the energy from the Sun, recycles water and other nutrients, and works with the electrical and magnetic forces to provide a moderate climate. The atmosphere also protects us from high-energy radiation and the frigid vacuum of space. Without water, the elixir of life, the biosphere that exists on the surface of the earth would not be possible. Water is vital for all living beings. The solid component of the upper part of earth is called the lithosphere. Soil is the uppermost layer of the lithosphere. The surface of the Earth (air, water and land) is called the biosphere, i.e. *Where Living Things Exist.* It is composed of smaller units called the ecosystems. A detailed discussion on the various components is given in the respective units.

Environmental degradation

Air and water pollution, soil degradation, deforestation, desertification, shrinking wetlands, inadequate public health and sanitation, water scarcity, falling groundwater tables, over extraction of water for irrigation purposes are some of the global environmental problems. Environmental degradation, poverty and economic development are closely inter-linked. Some of the key environmental problems that lead to degradation include the following:

- Rapid population growth
- Rapid consumption of resources
- Little emphasis on pollution prevention and waste reduction strategies
- Degradation of parts of the earth's life support systems
- Poverty, which drives poor people to use renewable resources unsustainably

- Failure of economic and political systems to encourage sustainable development
- Management of our environment with little knowledge about how ecosystems work

According to the model given by Paul Ehrlich and John Holdren (1972), a given area's total environmental degradation and pollution depends on three factors:

1. Number of people (population size P)
2. Average number of units of resources each person uses (per capita consumption or affluence A)
3. Amount of environmental degradation and pollution produced for each unit of resource used, T

Population × Affluence × Technology = Environmental Impact

The environmental problems facing India are different from those facing the affluent countries. Some major environmental problems in India are mentioned below.

- Land degradation, deforestation practices
- Air, soil, water pollution, hazardous wastes and improper solid waste disposal
- Unplanned power generation, shortage of electricity and power quality and quantity
- The loss of wildlife
- Destruction and erosion of genetic resources (crop, animal, fish, tree genetic resources)

Environmental Management System (EMS)

Environmental management and economic development are mutually supportive aspects of the same agenda. The ISO 14001 (International Organization for Standardization, Geneva) defines it "as *the overall management system that includes organizational structure, planning activities, responsibilities, practices, procedures, processes and resources for developing, implementing, achieving, reviewing and maintaining the environmental policy*". The ISO 14001 environmental management system standard is the most widely recognized EMS framework and more than 12,000 entities have had their EMSs certified conforming to ISO 14001.

Objectives of environmental management

1. To protect the environment
2. To promote the prudent use of natural resources
3. To increase awareness among residents, establishments, institutions, the general public of the value and importance of the environment
4. To actively promote environmental policies in all aspects of management and its implementation.

Some important Acts and Notifications relating to the environment

Various regulatory measures have been taken in our country for the preservation of our environment. The regulatory and institutional decision-making framework for environmental protection in India is embodied in the major acts approved by the Indian Parliament. Some of these acts are named below:

1. The Water Prevention and Control of Pollution Act of 1974, amended in 1991, which established the Central Pollution Control Board (CPCB) and State Pollution Control Boards (SPCB)
2. The Air Prevention and Control of Pollution Act of 1977 amended in 1981, 1988
3. The Noise Pollution Act, 1987
4. The Environment Protection Act of 1986
5. The Wildlife Protection Act, 1972, amended in 1983, 1986 and 1991
6. The Forest Conservation Act of 1980, amended in 1988
7. The Motor Vehicle Act of 1938, amended in 1988
8. The Public Liability Insurance Act of 1991

9. Notifications on the Coastal Regulation Zone, 1991
10. Environmental Impact Assessment of Development Projects, in 1994
11. The National Environment Appellate Authority Act, 1997.

The objectives of the above Acts and Notifications are aimed at strengthening India's environmental management system.

Environment Action Programme - India (EAP)

The Ministry of Environment and Forests through the government, in 1993, issued the Environment Action Programme - India (EAP). The EAP identified seven environmental priorities for India. They include the following:

1. Conservation and sustainable utilization of biodiversity in selected ecosystems including forests, mangroves, wetland, coral reefs, mountain ecosystems
2. Afforestation, wastelands development and conservation of soil and ensuring that water sources are not polluted
3. Control of industrial pollution, reduction and management of wastes
4. Improving access to clean technologies
5. Tackling urban environmental issues
6. Alternative energy plan
7. Strengthening scientific understanding of environmental issues and creating environmental awareness.

The Ministry of Environment and Forests is responsible for planning, promoting and coordinating the implementation of various environmental and forestry programs. The objectives are supported by a set of legislative and regulatory measures aiming at preservation, conservation and protection of the environment. The ministry has identified a set of targets to be achieved to meet the objectives of the Tenth Plan for the period 2002-2007:

1. Increase in forest and tree cover to 25 per cent by 2007 and 33 per cent by 2012
2. All villages to have sustained access to potable drinking water within the Plan period.
3. Cleaning of major polluted rivers by 2007 and other notified stretches by 2012
4. Environmental management and economic development
5. To promote sustainable development
6. To mainstream environmental issues which will include the incorporation of environment in the curriculum, sharpening the mechanism for the environmental auditing of the development activities, creating heightened awareness, broad basing community involvement and to forging a coordinated effort with corporate and public sector in environmental management
7. Increasing productivity
8. Efficient management of resources
9. Focus on the local needs and location specific strategies in terms of species selection and management of resources.

Some important organizations in India working in the field of environment

There are several leading NGO's in India working on different issues in the field of environment where they have developed expertise. Centre for Science and Environment is one of India's leading environmental NGO with a deep interest in sustainable natural resource management and they bring out a bimonthly magazine, *Down to earth* where current environmental issues are brought to the notice of the general public. TERI is an organization based in New Delhi. Their main focus is on efficient utilization of energy, sustainable use of natural resources, large-scale adoption of renewable energy technologies and the reduction of all forms of waste working for development towards sustainability. Some of the other important government organizations devoted to environmental management in India are the Geological Survey of India, Botanical Survey of India, Zoological Survey of India, Forest Survey of India, Wildlife Institute of India (Dehradun), Environmental Protection Agency, Forest Research Institute (Dehradun), Bombay Natural History Society (Mumbai) that bring out their magazine *Hornbill*, Salim Ali Institute for Ornithology and Natural History (Coimbatore), etc.

Solutions to global environmental problems require both a through awareness and knowledge of the problems. We have a moral obligation to keep our environment clean and free from pollution and hand it over safely to our future generations.

The Environment Calendar

World Wetland Day	February 2
World Forestry Day	March 21
World Water Day	March 22
World Meteorological Day	March 23
Earth Day	April 22
Anti Tobacco Day	May 31
World Environment Day	June 5
World Population Day	July 11
Ozone Day	September 16
Green Consumer Day	September 28
World Habitat Day	October 3
World Animal Welfare Day	October 4
World Habitat Day	October 4
Wildlife Week India	October 1 – 7

International Day for Natural Disaster Reduction	October 13
International Day for biological Diversity	December 29

Some Interesting Facts

- French fries and Potato chips contain high amount of the chemical acrylamide. Acrylamide is a carcinogen and is formed when foods high in starch content are subjected to high temperatures.
- Aspartame, an artificial sweetener used in processed foods, sugar free tablets, ice creams, soups, custard powders etc. is a carcinogen. This was the finding of the European Foundation of Oncology and Environmental Sciences (EFOES), Bologna, Italy. It is toxic and causes arthritis, asthma, brain cancer, diarrhea, hypertension, and memory and vision loss.
- The total number of chemicals known to exist are more than 7 million, of which 1,00,000 is commonly used. Around 1500 new chemicals are introduced annually in the world without or with rudimentary tests for safety and are introduced into cosmetics, toys, household appliances etc.
- New Delhi tops cadmium content and Mumbai tops lead content in toys. In a recent study conducted, majority of toys manufactured had lead, cadmium and polyvinyl chloride (PVC). To protect the PVC from degradation due to exposure to air, lead and cadmium are added as stabilizers that leach out every time it is exposed to light or being chewed by the child. This is detrimental to the health of children and is known to cause retarded mental development, kidney malfunction etc.
- Several soy and herbal medicines may increase the risk of breast cancer. The ingredients used in these preparations have adverse effects and lead to various health disorders.
- The Maharashtra Food and Drug Administration banned the sale of Rajnigandha – a form of gutkha or pan masala. It contains magnesium carbonate, a carcinogen.

- The Food and Drug Administration, Mumbai detected light liquid paraffin oil in some of the products manufactured by Johnson and Johnson Ltd. This oil prevents the baby's skin from breathing.
- The European Commission has banned 22 chemicals used in hair dyes. This bans all temporary and permanent hair dyes that are not safe and it would come into effect from 1st December 2006.
- Despite DDT being banned in China, the breast milk of mothers from Hong Kong had high levels of the pesticide. This non-persistent organic pollutant is absorbed by infants from the breast milk.
- Approximately 67 million pounds of pesticides and 3 million tons of fertilizers are used annually on lawns in the US.
- India is one of the biggest exporters of leather but the European Nations had banned Indian leather in the early 1990's. Leather is treated with 20 – 40 different kinds of chemicals in the tanning processes. These include pentachlorophenol (PCP), chromium 6 and azo dyes that are carcinogenic to the skin. India has no certification process and no environment laws that make the certification of leather mandatory to ensure that hazardous and / or banned chemicals are not present in leather.
- The (EPA) Environmental Protection Agency, US bans Lindane, an organochlorine pesticide. Also lindane is used in lotions / shampoos in treatments for headlice and scabies.
- 40 % of the world's population lives in just two countries. They are India and China.

UNIT 2

ECOLOGY

Our planet Earth is a lovely beautiful white and blue ball when seen from space. It is the third planet from the Sun and it is the largest of the inner planets. It is the only planet known to support life and to contain liquid water at the surface. Our environment includes the biotic (living beings) and abiotic factors (like soil, water, temperature and light). The various segments of the environment consist of the atmosphere, lithosphere, biosphere and the hydrosphere. The surface of the Earth (air, water and land) is called the biosphere. It is composed of smaller units called the ecosystems. An ecosystem includes all the organisms and the non-living environment found in a particular place. They can be large or small. Our Earth hosts a variety of living organisms and the surface of the Earth, as a whole is an ecosystem. Each organism depends in some way on other living and nonliving things in its environment. The study of natural ecosystems will help us to understand the interrelations between living beings and the environment and how the impact of human beings is influencing the natural world.

Ecological case studies

1. Dwindling biodiversity in the Sariska Tiger Reserve, Rajasthan:

Poaching of the wild animals like leopards, tigers and other animals for skin, fur, natural resources, etc. in Sariska coupled with high amount of mining related problems, as the area is rich in dolomite, limestone, schist, marble and quartzite have led to severe ecological problems. There are over 200 mines in the protected forest area and more than 40 mines in the partly protected areas. The Sariska reserve, amidst the Aravallis, covers over 800 sq km of dry tropical forests. 'Dhok' is the principal tree species that covers 90% of the entire area and there are a variety of grasses and shrubs. This reserve supports over 100 species of birds and has the largest population of peafowl in the entire country. Leopards, Caracals, Sambar, Nilgai, Chital, Wild boar are found in this sanctuary. A rare feature of the fauna is the four horned antelope. Eco-development programs have just been implemented in Sariska with the formation of forest protection societies in villages and NGO's. The future of Sariska reserve lies directly in protecting and reducing the high biotic pressure that affects this unique Aravalli habitat of the Indian tigers.

2. Declining biodiversity in the wonderful Australian Woodlands:

It is predicted that over the next 50 to 100 years there will be a wave of bird extinctions paralleling the loss of many Australian mammal species witnessed earlier this century. The reason for bird species decline is due to practices such as hunting, predation or competition from introduced species and due to loss of habitat. The present case study is related to temperate woodland ecosystems and the birds dependent on them for habitat. Impacts on bird populations provide typical examples of the indirect pressures on biodiversity. Woodlands are distinguished from coastal forests and semi-arid lands by less marked seasonal variations in temperature, lower extremes of rainfall and a longer growing season. Woodland birds have evolved in response to the year-round availability of food and the variety of habitat features such as tree hollows, peeling bark, nectar and other carbohydrates provided by the predominant eucalyptus plant species. The year-round supply of food also means the forests are linked to surrounding habitats, as they are suitable for migratory birds, breeding or when food is scarce elsewhere in times of drought. More than 25% of land bird species found in woodlands are in decline or threatened. The paradise parrot and star finch are extinct. Endangered species include Regent honeyeaters, bush stone-curlew, squatter pigeon, superb parrot, swift parrot, turquoise parrot, grey-crown babbler, painted honey-eater and black-throated finch. In total, 16 woodland-dependent or woodland-associated species are threatened and many more are in decline, as concepts of conservation of ecosystems have not been scientifically followed.

Components of ecology

The term ecology is derived from two Greek words, *oikos* meaning dwelling place and *logos* meaning study. So ecology means the study of organisms in their home. The term ecology was termed by the German biologist Ernst Haeckel (1868). He defined ecology as the *'surrounding outer world'*, now called the environment. E.P.Odum (1969) defined the term ecology as the study of interrelationships between organisms and environment. Ecology is the study of interactions between organisms (biotic part) and their nonliving environment (abiotic factors). Ecology involves collecting information about organisms and their environment, looking for patterns, and seeking to explain these patterns.

Habitat is the place a plant or animal lives, while its *niche* is the specific place it occupies in an ecosystem. Thus, an ecosystem is the living organisms and the chemical and physical environment and ecology is the study of ecosystems. Living things are organized in an ecosystem by how they secure their food: autotrophs or heterotrophs. A very important factor determining the number and type of organisms living in an ecosystem is the amount of energy available. The amount of energy an ecosystem receives and the amount that is transferred from organism to organism has an important effect on the ecosystem's structure. All organisms need energy to carry out essential functions, such as growth, movement, maintenance and repair, and reproduction. Energy in an ecosystem flows from the sun to autotrophs (producers) then to organisms that eat the autotrophs, then to organisms that feed on other organisms. The primary source of energy for an ecosystem is the sun.

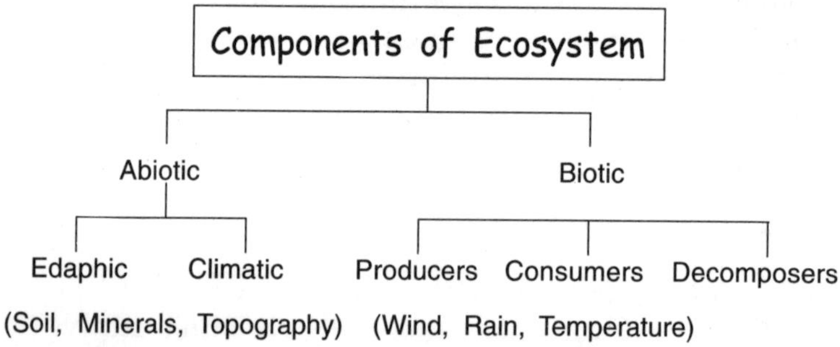

Ecosystem

Ecosystems are ecological units that include all the biotic factors and abiotic factors in an area. E. P. Odum defined ecosystem as the basic functional unit of organisms and their environment, interacting with each other and within their own components. Nearly all ecosystems are dependent upon the flow of solar energy and finite pools of nutrients. Carbon, hydrogen, nitrogen, oxygen, phosphorus and sulphur make up nearly all life on earth. The diversity of an ecosystem is a measure of the number of different species there, and how common each species is. Ecosystems are very complex and they can contain hundreds or even thousands of interacting species.

Environment and eco-factors

The structural and functional features, their size and composition are important characteristics of any ecosystem. The structural features include the biotic and abiotic components and the functional features include the functions performed i.e. food chains, webs, trophic levels, the flow of energy in the ecosystem, productivity, biogeochemical cycles and ecosystem regulation.

Structure of ecosystem

1. **Abiotic factors**

   They characterize the physical and chemical characteristics of the environment like the amount of sunlight, oxygen, CO_2, water, temperature, humidity, pH, and the availability of nutrients. They determine the type of organisms that can successfully live in a particular area.

   i. The Physical factors:

   - **Sunlight** is necessary for photosynthesis
   - **Water** is required by all living beings
   - **Temperature** all living things have range of temperatures for their survival
   - **Oxygen** it is necessary for cellular respiration
   - **Soil** the soil type, pH, the amount of water it holds, the available nutrients, etc determine the type of organisms that can successfully live in or on the soil (for example, cacti live in sand, cattails in soil saturated with water)

   ii. The Chemical factors:

   The availability of major trace elements, salts etc. can significantly affect the functioning of an ecosystem.

2. Biotic factors

They include all the living things that affect an organism. They include the plants, animals, fungi, bacteria and other living organisms inhabiting the area. Biotic components are often categorized as producers, consumers, and decomposers.

i. Autotrophs

They are autotrophic organisms with the capability of carrying out photosynthesis, and making food for themselves and storing some for other organisms to consume. Examples are plants, some bacteria, and algae. Autotrophs capture energy (i.e. sunlight) and incorporate it into organic compounds; therefore they are also called producers. Producers, such as plants with the green pigment chlorophyll, make food through a process called photosynthesis. In photosynthesis, plants use carbon dioxide and water to make sugar. The rate of photosynthesis is influenced by light intensity, temperature, and the availability of water. This food is used by the plant for its own energy or may be eaten by consumers. The types of autotrophs are:

- **Chemoautotrophs** are bacteria that obtain energy from oxidation of inorganic compounds such as ammonia, nitrites, and sulfides; they synthesize carbohydrates and are found in cave communities and ocean depths.
- **Photoautotrophs** possess chlorophyll and carry on photosynthesis. Algae are the main photoautotrophs in freshwater and marine environments. Green plants are the main land photosynthesizers.

ii. Heterotrophs

All other organisms in ecosystem are consumers or heterotrophs that eat the producers or other organisms. They eat the food that autotrophs have produced. The primary consumer is the herbivore. The secondary consumer eats the herbivores and the tertiary consumer feeds on secondary carnivores. Heterotrophs can be classified into different groups depending on what they eat:

- **Herbivores** are animals that feed directly on green plants.
- **Carnivores** are animals that eat other animals. The sequence of carnivores that feed in a chain can be referred to as primary, secondary and tertiary consumers.
- **Omnivores** are consumers that eat both plants and animals.
- **Decomposers** are heterotrophs that break down dead organic, the remains of dead plants and animals and waste products. They play a very important role in the ecosystem because they recycle nutrients where organic matter is converted into inorganic compounds like carbon dioxide. Bacteria and fungi are decomposers. They are also known as detritivores and feed on the "Garbage" of an ecosystem. Examples are microorganisms like bacteria, fungi and larger organisms like earthworms.

Functions of ecosystem

The ecosystems depend on solar energy to perform its function and this energy flows in the ecosystem, which is mediated through the processes explained in next page.

Trophic Structure Levels: Trophic levels are the feeding positions in a food chain such as primary producers, herbivores, primary carnivore, etc. The amount of living matter at a given point of time at each level is also called the standing biomass. Green plants form the first trophic level, the producers. Herbivores form the second trophic level, while carnivores form the third and even the fourth trophic levels. In an ecosystem these producers and consumers are placed in trophic levels or feeding levels. A trophic level is a group of organisms whose feeding source is the same number of steps from the sun. Most animals (carnivores) feed at more than one trophic level.

- Producers (autotrophs) occupy the first trophic level
- Herbivores occupy the second trophic level
- Carnivores occupy the third, forth and fifth trophic levels

Food Chain: A food chain indicates who eats whom in an ecosystem. The feeding of one organism upon another in a sequence of food transfers is known as a food chain. Another definition is the chain of transfer of energy from one organism to another. A simple food chain is like the following:

Rose plant → Aphids → Beetle → Chameleon → Hawk

Each organism occupies a separate trophic level. In this food chain, the rose plant is the primary producer (trophic level I). The aphids are the primary consumers (trophic level II) because they suck the juice from the rose plant. The beetle is the primary carnivore (trophic level III) because it eats the aphids. The chameleon, a secondary carnivore (trophic level IV), eats the beetle. The hawk is the tertiary carnivore (trophic level V) because it eats the secondary carnivore, the chameleon. The hawk eventually dies and decay-causing microorganisms break down the remains. Except in deep-sea hydrothermal ecosystems, all food chains start with photosynthesis and will end with decay. So the process of eating and being eaten forms a food chain. In nature, there are two general types of food chains. They are the grazing food chains and the detrital food chain.

1. **Grazing food chains**: It begins with leaves, stems and seeds or the green plants (producers), which are eaten by the herbivores, carnivores and the omnivores.
2. **Detrital food chain**: It begins with detritus or the dead organic matter, followed by detritiores, decomposers (including bacteria and fungi). Detrital food chains are connected to a grazing food chain when consumers of a grazing food chain feed on the decomposers of the detrital food chain.

Examples of some grazing food chains in different ecosystems

Forest ecosystem:

Trees ⟶ Aphids ⟶ Spiders ⟶ Small birds ⟶ Hawk

Cropland ecosystem:

Crop plants ⟶ Rodents ⟶ Snake ⟶ Hawk

Grass land ecosystem:

Grasses ⟶ Cattle ⟶ Lion

Aquatic ecosystem:

Phytoplankton ⟶ Zooplankton ⟶ Small fish ⟶ Large fishes

Example of a detrital food chain

Forest ecosystem:

Leaf litter ⟶ Algae, Bacteria ⟶ Copepods, Crabs ⟶ Small fishes ⟶ Fish eating birds

Food Web: A food web may contain several food chains. Energy flows through an ecosystem in a complex network of feeding relationships called a food web. A food web is defined as the relationship in which a predator eats different types of food at each trophic level and it is eaten by many different organisms. Several food chains involve a food web. It is the outcome of several complex-feeding interrelationships that exist in nature. For example in a grassland ecosystem, in the absence of a mouse, the grasses can also

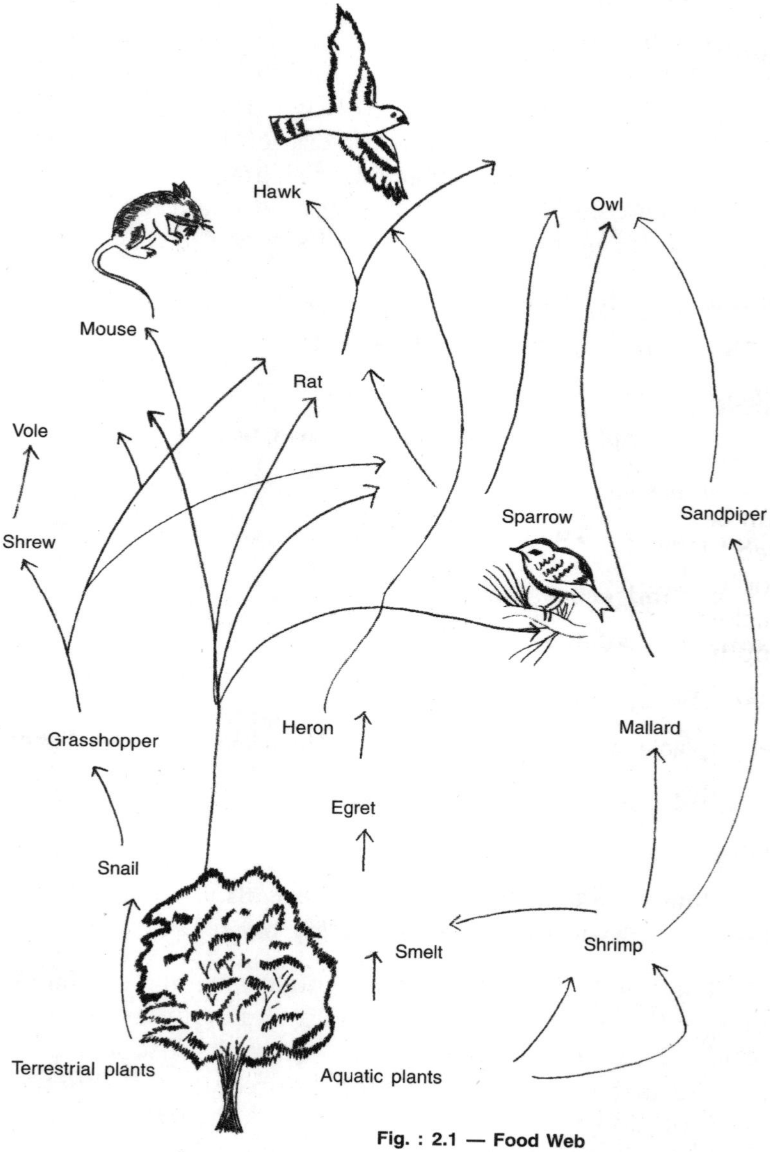

Fig. : 2.1 — Food Web

be eaten by rabbit, which can be eaten by a hawk. Again grasses can also be eaten by a grasshopper, which is preyed upon by a lizard and then eaten by the hawk. So in nature food chains are interlocked forming food webs (Fig. 2.1).

Significance of the food chain and webs

Food webs are important in maintaining the ecological balance and stability of ecosystems. For example, in grasslands, a decrease in the number of rabbits would lead to an increase in the mouse population etc. This would upset the ecological balance. A proper balance is important for the survival of all living beings in the ecosystem. Again they are important for biogeochemical cycles or nutrient cycles and energy flow through systems.

Ecological Pyramids

An ecological pyramid shows the trophic structure of an ecosystem as a graph representing the biomass, organism number, or energy content of each trophic level in a food web. The base of the pyramid represents the producer trophic level, and from there the consumer trophic level is stacked, with the apex representing the highest consumer trophic level. There are three types of ecological pyramids.

1. **Pyramid of numbers** is based on the number of organisms in each trophic level. It may be upright or inverted. In a grassland ecosystem, the producers are the grasses, which form a large base and are the maximum in number. They are followed by rabbits, mice (herbivores), then the snakes, lizards (secondary consumers), followed by the hawks (tertiary consumers) which gradually decrease in number forming a conical upright pyramid (Fig. 2.2).

   However, in a parasitic food chain, the pyramid could be inverted. The trees form the producers, which are less in number. They support many fruit eating herbivores, birds that in turn support a large number of parasites and hyperparasites. In this case, the

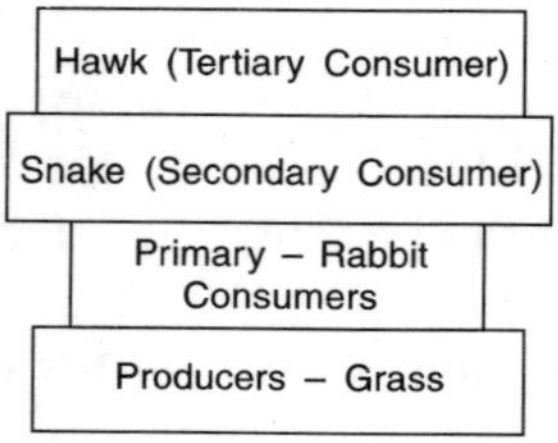

Grassland Ecosystem

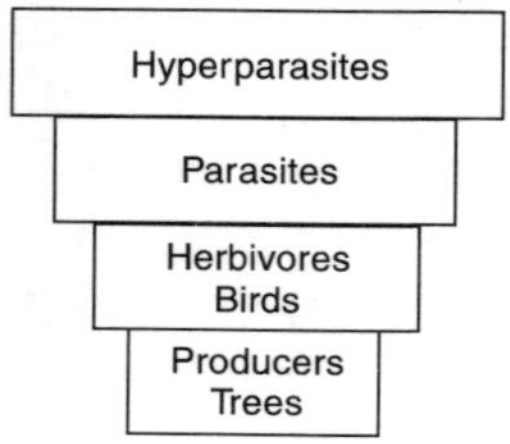

Fig. : 2.2 — Pyramid of numbers

pyramid is inverted as the number of organisms at each level increases to the top.

2. **Pyramid of biomass** is based on the weight (biomass) of organisms at each trophic level at one time. Usually a large mass of plants supports a medium mass of herbivores and a small mass of carnivores. Grassland and forest ecosystems show an upright pyramid. At one point in time at seashores, herbivores can have greater biomass feeding on algae that reproduce fast but are eaten, producing an inverted pyramid. Also, pond ecosystems show an inverted pyramid of biomass (Fig. 2.3).

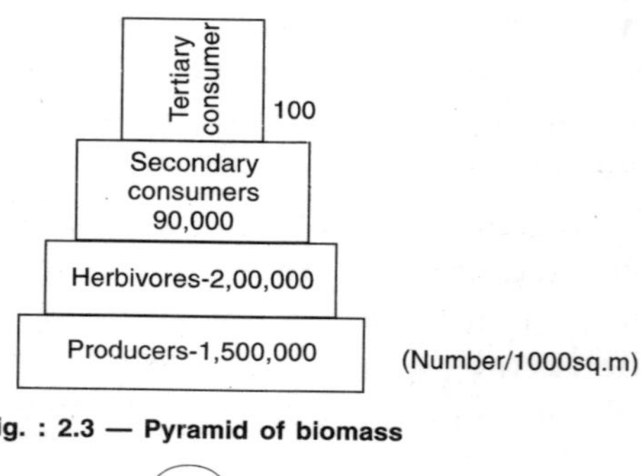

Fig. : 2.3 — Pyramid of biomass

3. **Pyramid of energy** is based on the total amount of energy present in each trophic level; it is always pyramidal. It means that assimilation and growth are not 100 % efficient. There is always a loss of energy at each level. Thus, the energy is highest at the producer level. About 10% of energy at a particular trophic level is incorporated into the next trophic level. Thus, 1,000 kg (or kcal in an energy pyramid) of plant material converts to 100 kg of herbivore tissue, which converts to 10 kg of first carnivores, which can support 1 kg of second level carnivores. This rapid loss of energy is the reason food chains have from three to four links, rarely five. This rapid loss of energy is also the reason there are few large carnivores (Fig. 2.4).

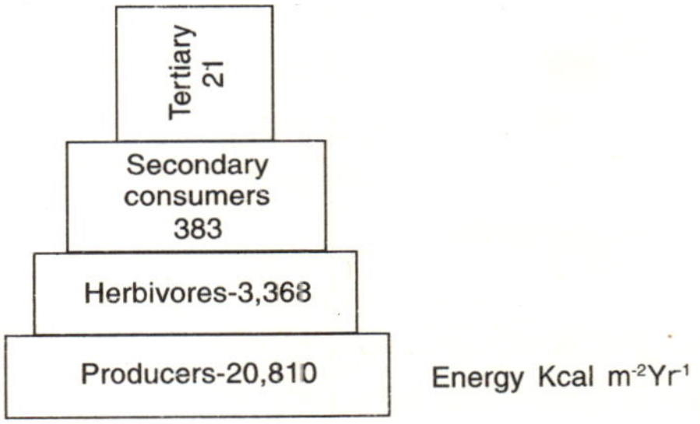

Fig. : 2.4 — Pyramid of energy

Flow of Energy in Ecosystem

Energy flows in an ecosystem through the food chains and food webs. The existence of life depends on the material cycling through the ecosystems. The sun and the solar energy is the driving force. All of the solar energy that enters an ecosystem is eventually lost as heat. Energy is lost or used as it flows through the trophic levels of an ecosystem. Producers absorb energy from the sun, but only about 50% of the energy captured from the sun becomes part of the plant's body. The other 50% is used for living and growing or lost as heat. At each trophic level, the energy stored in an organism is about 1/10. Since energy diminishes at each trophic level, few ecosystems contain more than 4 or 5 trophic levels. The energy flow is a consequence of two fundamental laws of thermodynamics.

First law of thermodynamics: It states energy can neither be created nor destroyed; it can only be changed from one form of energy to another. For example, the green plants (producers) containing the pigment chlorophyll capture solar energy. This is transformed into biochemical energy in the plants. So this way energy is transformed or changed from one form to another.

Second law of thermodynamics: When energy is transformed from one form to another, there is always some loss of energy from the system, usually as low-grade heat. Energy is converted from a more useful to a less useful or disorganized form. So energy is degraded and this disorganization of energy is called entropy. Energy is maximum at the I trophic level or in the producers. There is loss of energy at each trophic level, through the processes of respiration, locomotion and other energy consuming activities.

Ecological Balance and Consequences of Change

The ecosystems undergo changes over the geological time and the system automatically resists the changes and comes back into equilibrium by a mechanism known as homeostasis. Feedback is very important in ecosystems. There are two feed backs, the positive feed back which results in decline and collapse of the ecosystem and the negative feed back which results in homeostasis or equilibrium.

When we start to disturb this balance for our needs we introduce an element of imbalance into its working. An example is cited to explain the same. Deforestation practices have destroyed the trees, and it has caused the disappearance of the deers, tigers, beetles, birds, lichens, mosses, mushrooms and a host of other forms of the living world. Consequently the up take of soil water and its release into the atmosphere through transpiration is affected. The rainfall is affected and subtle changes in the climate set have in. Nature works in ecosystems and an ecosystem comprises of a group of living organisms and their physico-chemical environment. A system is an entity in its own right and a disturbance at one point is sure to have repercussions all over the system. If not repaired on time, the system is liable to disintegrate. In every ecosystem, there is a naturally attained balance in which each component is in perfect harmony with the rest. When man clears a forest or builds a dam, he interferes with the ecological balance that has been there. Without intending it, he disturbs the ecological balance of the area and is seeking to establish a new ecosystem. A new system can come only in the place of an old one. In the bargain, the inhabitants of the old system disappear. When older ecosystems are upset over a large area, the migratory, the adaptive and such other limitations cause the total disappearance of a few species. This is undesirable not only because of the richness of the flora and fauna suffer but more so because in the new dispensation, there might be disastrous consequences. We are part and parcel of a natural ecosystem and our survival depends very much on the preservation of this ecosystem. If we disturb this ecological balance that nature worked out through ages we are at our own risk.

Biogeochemical Cycles

The Earth is essentially a closed system with respect to matter, and all matter on the Earth cycles. The biogeochemical cycle is the movement or cycling of matter through a system (through the atmosphere, hydrosphere, lithosphere and biosphere). Biogeochemical cycles are the circulation pathways of elements (e.g., carbon, oxygen, hydrogen, nitrogen or mineral elements) through the biotic and abiotic components of an ecosystem. In addition to energy cycling, the nutrient cycling is an important aspect in ecosystems. A reservoir is that portion of the earth that acts as a storehouse for the element. An exchange pool is the portion of the environment from which producers take chemicals, such as the atmosphere or soil. The biotic community is the pathway (i.e., food chains) through which chemicals move. Energy flows through an ecosystem and nutrients cycle in an ecosystem. The main cycles are the water cycle, carbon cycle, nitrogen cycle, sulphur cycle, oxygen cycle and the phosphorous cycle. Some important cycles are discussed below.

Carbon Cycle

Photosynthesis and respiration form the basis of this cycle (Fig. 2.5). The Earth's atmosphere contains carbon dioxide and during photosynthesis, plants use CO_2 along with water and solar energy, for making organic compounds (carbohydrates).

$$6CO_2 + 6H_2O \xrightarrow{\text{sunlight chlorophyll}} C_6H_{12}O_6 + 6O_2$$

They store the carbon for themselves and other organisms. The autotrophs and heterotrophs use oxygen in cellular respiration to break down carbohydrates. The consumers obtain energy molecules containing carbon by feeding upon plants and animals.

During respiration, the living organisms break down complex organic molecules, returning carbon dioxide and water.

$$C_6H_{12}O_6 + 6O_2 \longrightarrow 6CO_2 + 6H_2O + \text{energy}$$

Carbon thus continuously moves from the atmosphere into the food chain during photosynthesis and returns back to the atmosphere during respiration.

Respiration by living organisms including decomposers feeding on the dead organic matter return carbon dioxide either to the oceans or the atmosphere. A very small portion of the dead organic matter ends up being buried in the sediments resulting in the formation of fossil fuels. When these are burnt, carbon in the form of carbon dioxide is returned to the atmosphere.

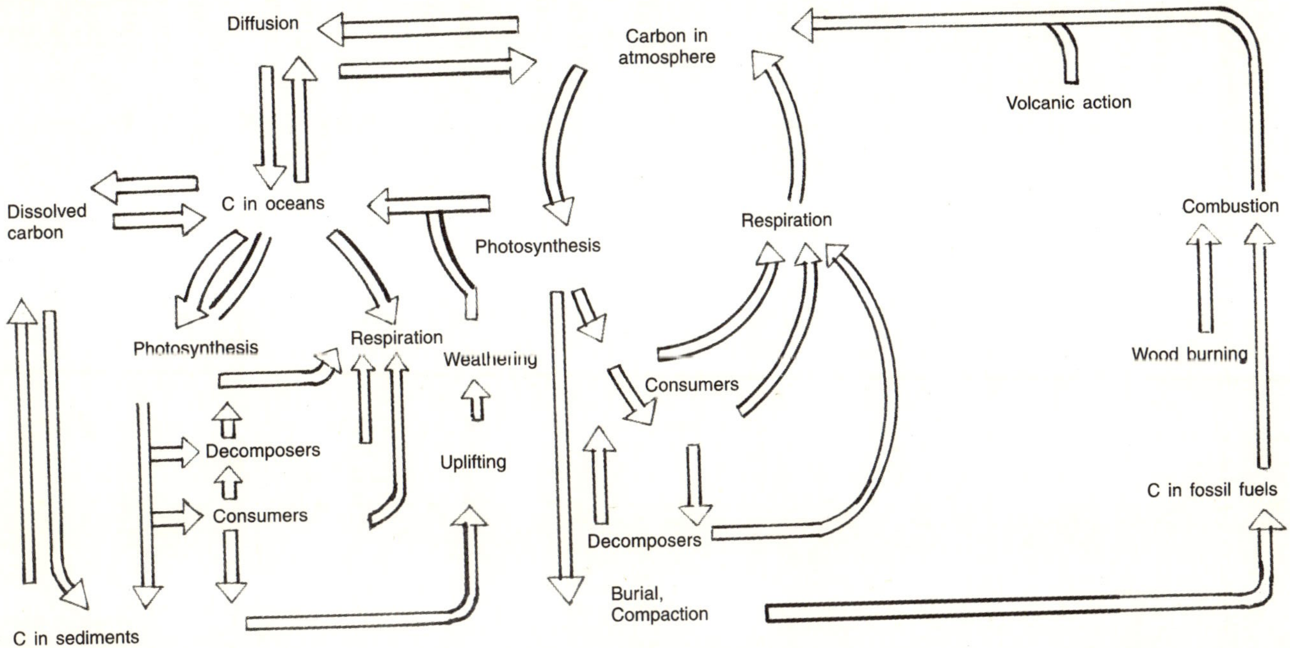

Fig. : 2.5 — **Carbon Cycle**

Nitrogen Cycle

Though 78% of the atmosphere is comprised of free nitrogen, the major reservoirs do not contain nitrogen in easily available form for the plants. As Plants cannot incorporate N_2 into organic compounds and they depend on microorganisms to make nitrogen available to them. The sequence of transformations from free atmospheric nitrogen to inorganic fixed nitrogen, to simpler organic compounds, to complex organic compounds stored in the tissues of animals and plants and eventually in the release of this nitrogen back to the atmosphere is known as the nitrogen cycle. It comprises of the following steps (Fig. 2.6).

Nitrogen fixation: Some microorganisms (*Clostridium, Azotobacter*) living in the root nodules of leguminous plants convert the atmospheric nitrogen to fixed nitrogen which

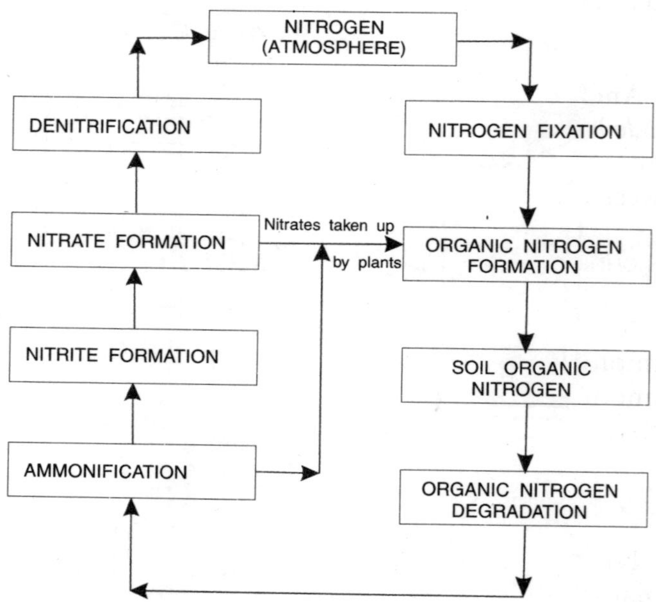

Fig. : 2.6 — Nitrogen Cycle

is available to plants for use. The overall biochemical reaction for nitrogen fixation is as follows:

$$N_2 + 6e^- + 12ATP + 12H_2O \xrightarrow{\text{nitrogenase complex}} 2NH_4^+ + 12ADP + 12P_i + 4H^+$$

Organic nitrogen formation: The nitrogen thus available to the plant is converted into plant proteins. When animals and higher organisms consume plants they are converted into animal proteins.

Soil organic nitrogen: Excretory products of animals, plant tissues, dead and decaying plant and animal matter and microorganisms form the soil organic nitrogen.

Organic nitrogen degradation or proteolysis: The soil organic nitrogen contains nitrogen in plant and animal proteins that are organically bound. The nitrogen is set free by microbially aided (*Clostridium* species) enzymatic hydrolysis of the proteins into simpler amino acids.

$$\text{Proteins} \xrightarrow{\text{proteinases}} \text{Peptides} \xrightarrow{\text{peptidases}} \text{Amino acids}$$

Ammonification: Bacteria like *Clostridium, Proteus* and *Pseudomonas* species are able to carry out the above reaction. Here amino acids are transformed into ammonia, the process commonly known as ammonification.

$$CH_3CHNH_2COOH + 1/2O_2 \xrightarrow{\text{alanine deaminase}} \underset{\text{(Pyruvic acid)}}{CH_3COCOOH} + \underset{\text{(Ammonia)}}{NH_3}$$

Nitrite formation: Microorganisms transform the ammonia to nitrate by the process nitrification. First ammonia is oxidized to nitrite by ammonia oxidizing bacteria (*Nitrosomonas europaea, Nitrosovibrio tenuis*).

$$\underset{\text{(Nitrite)}}{2NH_3} + 3O_2 \rightarrow 2HNO_2 + \underset{\text{(Nitrate)}}{2H_2O}$$

Nitrate formation: In the second step nitrite is transformed into nitrate by nitrite oxidizing bacteria (*Nitrobacter winogradskyi, Nitrospina gracilis*).

$$\underset{\text{(Nitrite)}}{HNO_2} + 1/2O_2 \rightarrow \underset{\text{(Nitrate)}}{HNO_3}$$

Denitrification: Nitrates serve as plant food and are converted to plant proteins. The next step is denitrification, where nitrates are converted into gaseous nitrogen. This reaction is carried out by several bacterial species (*Achromobacter, Thiobacillus, Vibrio*).

$$\underset{\text{(Nitrate)}}{2NO_3^-} \rightarrow \underset{\text{(Nitrite)}}{2NO_2^-} \rightarrow \underset{\text{(Nitric oxide)}}{2NO} \rightarrow \underset{\text{(Nitrous oxide)}}{N_2O} \rightarrow \underset{\text{(Nitrogen)}}{N_2}$$

Environmental Impact Assessment (EIA)

The objectives of environmental management should be to achieve the greatest benefit presently possible for the use of natural resources without reducing their potential to meet future needs and the carrying capacity of the environment.

Environmental impact assessment is a tool used for decision-making regarding projects, developments and programs for example incinerators, control technologies, airport runways, etc. EIA identifies the environmental, social and economic impacts of a proposed development prior to decision-making. EIA means an examination, analysis and assessment of planned activities with a view to ensure environmentally sound and sustainable development.

The objective of the EIA is to ensure that environmental aspects are addressed and the potential problems are foreseen at the appropriate stage of project design. EIA should be envisaged as an integral part of the planning process and initiated at the project level from the start.

Principles of Environmental Impact Assessment

The main principles of environmental impact assessment include the following:

Participation and transparency

It should include adequate participation of all stakeholders. EIA should be conducted through an established process and it should be clearly understood by all participants.

Certainty

It should have clear objectives, be consistent, and be conducted within fixed time-frames.

Integrity

All decisions should be based on the best available information, and all relevant factors need to be taken into account by decision-makers. Where impacts are uncertain, outcomes should rely on sound risk assessment and management.

Accountability and cost-effectiveness

The decision makers within the government should be able to provide clear and detailed reasons for their decisions to all stakeholders. The EIA process should include: outline of the proposal, project design, construction, operation and finally decommissioning. The project operators should be accountable for commitments made during project approval. The process should meet its objectives while imposing the least cost to participants.

Flexibility and practicality

The process should be able to accommodate proposals varying in type, scope of impact, and complexity. Flexibility is desirable in terms of the form of EIA process, issues to be addressed, process time-frames, and degree of public participation. The process should recognize community concerns, commercial realities, best practice technology, and scientific uncertainties.

Precaution

Where there are threats of serious or irreversible environmental damage, lack of full scientific certainty should not be used as a reason for postponing measures to prevent environmental degradation.

Some Interesting Facts

- Poaching, illegal logging and ecological damage is shrinking the rainforests in Africa to 8000 square kilometers per year.
- The Wandering Albatross is the bird with the largest wingspan, with a wingspan of 11 ft. 10 inches.
- A research team from Conservation International, USA have found 52 new species in Papua New Guinea that include 24 new species of fishes, 20 new coral species and 8 new shrimp species.
- The discovery of the Bird's Head Seascape, off Papua province has staked claims of being the world's richest area of biodiversity. Scientists report that the discovery would provide important information about the rare new ecosystem.
- The Amazon River accounts for one-fifth of the world's fresh water drained into the oceans. The Amazon basin occupies 7 million square kilometers and is a home of millions of species. The Amazon has the largest water lily in the world, the *Victoria Amazonica.* This plant has leaves that can attain a diameter of 6 ft. Temperate rainforests are one of the most endangered forest types in the world. They are more endangered than the tropical rainforests. They are one of the most productive ecosystems on our earth in terms of biomass and accumulate as much as 500 to 2,000 metric tons of organic matter per hectare.
- The western red cedars and yellow cedars on the west coast of North America can live up to about 2,000 years. A yellow cedar here is known to be 1,835 years old. The alerce tree in Chile lives up to aproximately 4,000 years. One alerce is known to have lived for more than 3,600 years, as calculated from its tree rings.
- The tallest trees in the world are the Giant Sequoia trees that make up the famous Redwood forests of California, USA.
- Lake Superior is the greatest Great Lake and the largest freshwater lake in the world by surface area, covers 31,700 square miles (82,100 square kilometers) and the third largest by volume. Only Lake Baikal in Siberia and Lake Tanganyika in East Africa contain a greater

volume of fresh water. The lake has a retention time (how long water stays in the lake, based on volume and the mean rate of outflow) of 191 years. It is the deepest of the Great Lakes and contains three quadrillion (3,000,000,000,000,000) gallons of water. That's ten percent of the world's fresh surface water.

- The coral reefs are the main source of proteins for over 1 billion people.
- 24% of the world's terrestrial systems have been converted to cultivation lands.
- The Indian team, at the Tocklai Experimental Station, Jorhat, Assam, developed the world's first tea pill, which could be chewed and had the same refreshing effect as a cup of tea.

UNIT 3

AIR POLLUTION AND CONTROL

Rapid industrialization has created hazards to health and the environment. As far as our country is concerned, the Indian chemical industries are the second largest in the globe. Almost all the Indian metropolitan cities are seriously polluted. More than 10,000 polluting units lie in Delhi. The Indraprastha Thermal power station in Delhi emits tons of fly ash and SO_2 everyday. The same is the case of the thermal power stations at Ennore and Basin Bridge, Chennai. Some of the major industries contributing to air pollution levels in Mumbai are Tata thermal power plant, Bharat refineries, Hindustan petroleum etc. The importance of environmental issues is evident from the eco-watch columns of our newspapers. Also TERI, New Delhi has made digitalized electronic road displays in the capital where one could see the emission levels of important air pollutants, the humidity, temperature levels etc.

Matter can be in solid, liquid, or gaseous state and the same concept applies to air contaminants. Many metals are present in the atmosphere at trace levels. For example, particulate matter emitted from combustion of fossil fuels contains trace metals that were present in the original fuel sample. One of the greatest health hazards in air pollutants is from aerosols. Metals are emitted to the atmosphere from natural or anthropogenic sources (lead, beryllium, mercury, cadmium and chromium). Gaseous inorganic substances remain in the atmosphere for long residence times, and would continue to build up much higher levels in due course of time, if it were not for the gas phase reactions that convert these substances to water soluble species.

Atmosphere

The Atmosphere is defined as a thin envelope of air surrounding the earth. The present composition of the atmosphere is 78% nitrogen, 21% oxygen, and 1% comprised of other gases. Our current atmosphere is an oxidizing atmosphere, whereas the original atmosphere of the earth was a reducing atmosphere, as it did not contain oxygen. The original atmosphere may have been similar to the composition of the solar nebula and close to the present composition of the gas giant planets, though this depends on the details of how the planets condensed from the solar nebula. The earth's atmosphere 4.6 billion years ago was comprised of hydrogen and helium, the two most abundant gases found in the universe. Through the process of *out gassing*, i.e., the outpouring of gases from the earth's interior, many other gases were ejected into the atmosphere. These include water vapour, carbon dioxide and nitrogen. Plants, for example cyanobacteria, produced most of the oxygen of our atmosphere. As out gassing occurred over a period of millions of years, the atmosphere evolved to its present state. The atmosphere, solar energy, and our planet's magnetic fields support life on earth. The atmosphere absorbs the energy from the Sun, recycles water and other chemicals, and works with the electrical and magnetic forces to provide a moderate climate. The atmosphere also protects us from high-energy radiation and the frigid vacuum of space.

Composition of the atmosphere

The atmosphere is primarily composed of nitrogen (N_2, 78%), oxygen (O_2, 21%), and argon (Ar, 1%). A small percent of other very influential components are also present which include water, ozone, carbon dioxide. The pollution free atmospheric constituents are outlined below.

Table 3.1 : Major and minor constituents of the atmosphere		
Major constituents: (% / Volume)	Nitrogen (78.09) Oxygen (20.94) Water vapour (0.1 – 5)	N_2 O_2 H_2O
Minor constituents: (% / Volume)	Argon (9.34×10^{-1}) Carbon dioxide (3.25×10^{-2})	Ar CO_2

The trace constituents include neon, helium, methane, nitrous oxide, hydrogen, xenon, sulfur dioxide, ozone, nitrogen dioxide, ammonia, carbon monoxide, iodine and particles (dust, soot, etc).

Layers of the atmosphere

The envelope of gases surrounding our Earth changes from the ground towards the upper layers. Thermal

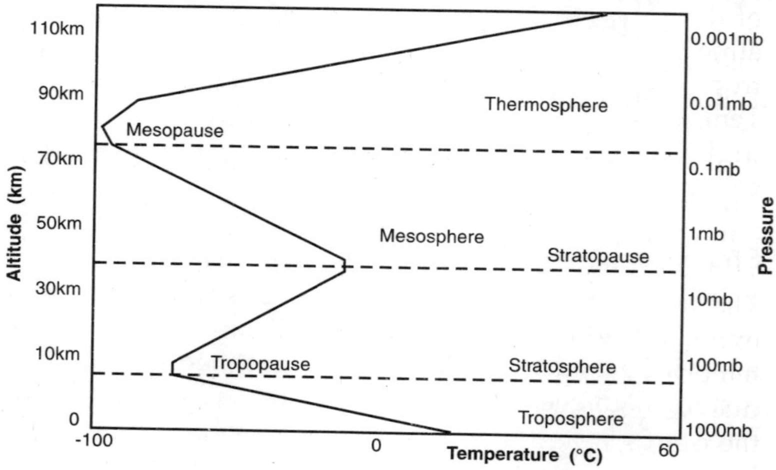

Fig. : 3.1 — Layers of the atmosphere

characteristics, chemical composition, movement, and density are the main criteria used to identify the four distinct layers of the atmosphere (Fig 3.1). The total mass of the atmosphere is about one millionth of the earth's total mass i.e. approximately 5×10^{15} tons. The atmospheric parameters vary with altitude, temperature and chemical species present as can be seen from the table below.

Table 3.2 : Temperature variations in different layers of the atmosphere

Atmospheric layers	Altitude (Km)	Approximate temperature (^{0}C)
Thermosphere	85 – 600	-93 to 1700
Mesosphere	50 – 85	-3 to –93
Stratosphere	11 – 50	-52 to –3
Troposphere	0 – 11	15 to –52

1. Troposphere

The troposphere starts at the Earth's surface and extends 0 to 11 kilometers high and is most dense of the atmosphere (contains 70% of the mass of the atmosphere). The troposphere is where all weather phenomena takes place and the air pressure at the top of the troposphere is only 10% of that at sea level (0.1 atmospheres). There is a thin buffer zone between the troposphere and the next layer called the tropopause. Temperature drops from about 15 to -56°C as we go higher. The tropopause and the troposphere are collectively known as the lower atmosphere.

2. Stratosphere

The stratosphere starts just above the troposphere and extends to 50 kilometers (31 miles) high. The temperature in this region increases gradually to -3°C due to the absorption of UV radiation. As compared to the troposphere, this part of the atmosphere is dry and less dense. The airflow is mostly horizontal and the ozone layer in the upper stratosphere has a high concentration of ozone, primarily responsible for absorbing the ultraviolet radiation from the Sun. The formation of this layer is a delicate matter, since it can form only when oxygen is produced in the atmosphere. There is considerable recent concern that man-made fluorocarbon compounds are depleting the ozone layer. 99 % of air is located in the troposphere and stratosphere. The stratopause separates the stratosphere from the next layer.

3. Mesosphere

The mesosphere starts just above the stratosphere and extends to 85 kilometers (53 miles) high. In this region, the temperatures again fall as low as -93°C as it increases in altitude. The mesopause separates the mesosphere from the thermosphere. The regions of the stratosphere and the mesosphere, along with the stratopause and mesopause, are called the middle atmosphere.

4. Thermosphere

The thermosphere starts just above the mesosphere and extends to 600 kilometers (372 miles) high. It is the warmest layer of the atmosphere. The temperatures go up as it increases in altitude due to the Sun's energy. Temperatures in this region can go as high as 1,727 °C. Chemical reactions occur much faster here than on the surface of the Earth. This layer is known as the upper atmosphere. The lower layer of thermosphere is called the Ionosphere.

Ionosphere

The ionosphere is very thin, but it is where aurora takes place, where atoms are ionized and it is also responsible for absorbing the most energetic photons from the Sun. It is an electrified region within the upper atmosphere. It contains large concentrations of ions and free electrons. Here the oxygen and nitric oxide occur in an ionized state and ultraviolet radiation causes their ionization. The ionosphere is very important for radio wave propagation and for reflecting radio waves, thereby making long-distance radio communication possible. The ionosphere for radio wave (AM only) propagation. It is composed of the D (60 Km), E (120 Km), and F (180 Km) layers, the D layer is good at absorbing AM radio waves and it disappears at night. The E and F layers bounce the waves back to the earth, which explains why radio stations adjust their power output at sunset and sunrise. We have learnt to use the properties of the ionosphere in many ways over the last century like radio, television, and satellite communications.

Exosphere

The exosphere starts at the top to the thermosphere and continues until it merges with interplanetary gases, or space. Hydrogen and Helium are the prime components and are only present at extremely low densities.

Earth's Energy Balance

The earth's energy balance and solar radiation is important in understanding global processes and environmental

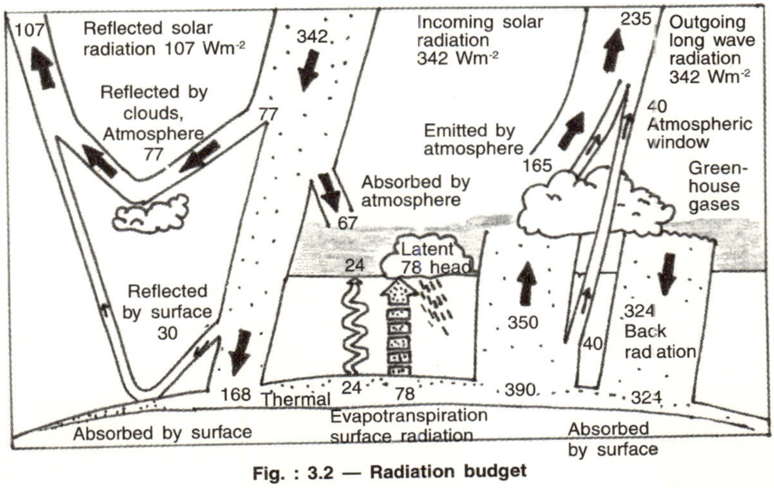

Fig. : 3.2 — Radiation budget

problems associated with the atmosphere such as the green house effect and ozone depletion. The earth receives energy from the sun and this affects the atmosphere, oceans, land and living beings before being radiated back into space. The earth's energy balance refers to the equilibrium between the incoming and outgoing energy, which involves a change in the energy's form (Fig. 3.2). This balance will change if the solar output or earths albedo changes. That has indeed happened in the past. The earth intercepts only a tiny fraction of the total energy emitted by the sun. This energy sustains life, drives many processes at or near the earths surface including the hydrological cycle, ocean waves and circulation of air on a global scale. Nearly, all of the energy that is available at the earth's surface comes from the sun. The internal energy of the earth is a very small fraction of the earth's energy budget. This internal heat drives the tectonic cycle, generates earthquakes and creates volcanoes.

The Earth's surface reflects a part of the solar energy. A significant part of the Earth's reflectivity can be attributed to clouds. The reflectivity of a planet is referred to as the albedo (albedo means reflectivity) and is expressed as a fraction. The albedo of Earth depends on the geographical location, surface properties, and the weather. On an average the earth's albedo is about 0.3. This fraction of incoming radiation is reflected back into space. Our planet absorbs the remaining 0.7 part of the incoming solar radiation.

On absorbing the incoming solar radiation, the Earth warms up, like a black body and it results in a rise in temperature. If the Earth had no atmosphere or ocean, as is the case on the moon, it would get very warm on the sunlit part of the planet and much colder than we experience presently, on the darker side. All heated objects emit electromagnetic radiation, particularly when they are surrounded by empty space. This radiation is referred to as outgoing radiation. As long as the incoming radiative flux is larger than the outgoing, the radiated object will continue to warm, and its temperature will continue to increase. This will result in an increase in the outgoing radiation (according to the Stefan-Boltzman law the outgoing radiation increases faster than the temperature). At some point, the object will emit as much radiation as the amount incoming and a radiative equilibrium or balance will be attained.

The effective temperature is the temperature at the surface of the planet if it had no atmosphere. The effective temperature of Earth is about 255 K (or -18 °C). With this temperature, the Earth's radiation will be centered on a wavelength of about 11 ¼m, well within the range of infrared (IR) radiation. Averaged over all seasons and the entire Earth, the surface temperature of our planet is about 288 K or 15°C. This difference is in the effect of the heat absorbing components of our atmosphere. This effect is known as the **greenhouse effect**, referring to the practice of warming garden plants by covering them with a glass enclosure. The atmosphere is almost transparent to sunlight and all the energy that is absorbed at the surface results in warming and the emission of IR radiation. This radiation cannot freely escape into space because of absorption in the atmosphere by trace gases such as water vapour and carbon dioxide. These absorbing gases and their surrounding air warm up, emitting radiation downward, towards the Earth's surface, as well as upward, towards space. This effectively traps a part of the IR radiation between the ground and the lower 10 km of the atmosphere. This reduction in the efficiency of the Earth to lose heat causes the surface temperature to rise above the effective temperature until finally, enough heat is able to escape to space to balance the incoming solar radiation. The greenhouse effect forced the planet to raise its surface temperature until the amount of heat radiated from the top of the absorbing layer is equal to the solar radiation at the top of the atmosphere. It is at the top of the absorbing layer that the effective temperature is reached, while down at the surface of the Earth it is much warmer.

Climate

Climate is the average weather pattern in a place for over some period of time (i.e. more than thirty years), including seasonal variations. When scientists talk of climate change, they refer to the variations of the earth's climate over a defined period of time. Therefore, climate is the statistical description of all weather conditions in a given location and the global climate is the statistical description of all weather conditions in the world. Regional climate is the climate at a given place in the world. The climate of a particular region depends on many factors. They are the amount of sunlight received, altitude, topography, and how close it is to the oceans. Climate varies with latitude as equatorial regions receive more sunlight than the Polar Regions. The global climate is defined as the climate of a planet as a whole, with all the regional variations averaged. Global climate depends on the amount of energy received by the sun and the amount of energy that is held in the system. These amounts vary for different planets. Weather can change in may be just a few hours, but the climatic changes take longer timeframes. The climatic events, like an El Nino, happen over several years. The Earth's global climate is a dynamic system driven by the amount of solar radiation, atmospheric chemistry, amount and types of clouds, and the influence of the biosphere. A change in the temperature results in a change in other factors that affect climate such as clouds or precipitation. The position of the earth with respect to the sun is related to the changes in climate that lead to the recession of the ice ages. Cyclic changes occur in the earth's orbit around the sun, the tilt of the earth's axis and the inclination of the axis with respect to its orbit. These changes in the earth's position occur very slowly and the corresponding changes in the climate are also slow. The terrestrial factors, which influence climate, include changes in oceans, composition of air and the reflectivity of the earth atmosphere system. The small-scale fluctuations happen over decades and larger climate changes happen over hundreds or even thousands of years.

Today, global warming is a serious environmental issue and our Earth is warming more quickly than it has in the past. Climate change has affected our earth for more than 4.6 billion years. It has repeatedly warmed and cooled and the changes have affected the types of environments on Earth's. The evidence of past climate change or paleoclimates can be seen from the ancient sediment deposits, fossils etc. Due to the global nature of the issue, international efforts are examining the implications of climate change. The Inter-Governmental Panel on Climate Change (IPCC) under the auspices of United Nations Environmental Program (UNEP) is examining the complex issue of climate change.

Weather

The weather of our earth is the state of the atmosphere at any given time and place. Weather takes place in the lower layer of the atmosphere, i.e. the troposphere. The important factors that determine weather are temperature, precipitation, fronts, clouds, and wind. Meteorology is the study of weather, and meteorologists are scientists who study and predict weather. Our atmosphere is in constant motion from the Sun's energy reaching the surface and this causes weather to occur. Examples of severe weather conditions are hurricanes, tornadoes and thunderstorms. High pressure occurs when the air is sinking. During these conditions a good and clear weather is seen because the air is stable. However, low-pressure areas are associated with clouds, storms and wind. This happens because air rises when it is unstable. Weather changes every season because of the Earth's tilt when it revolves around the sun.

Dispersion of Pollutants in the Atmosphere

In order to control air pollution and perform air quality monitoring, it is important to understand the principles of transport and dispersion of pollutants in air. The dispersion and transport of air pollutants are influenced by many factors. Global, regional weather patterns and local topographical conditions affect the way that pollutants are transported and dispersed. The primary factors affecting transport and dispersion of pollutants are wind, stability and turbulence. It is the natural horizontal motion of the atmosphere and occurs when warm air rises, and cool air comes in to take its place. It is caused by differences in pressure in the atmosphere. Pressure is the weight of the atmosphere at a given point. Atmospheric weight is determined by the height and temperature of a column of air. Since cool air weighs more than warm air, a high-pressure mass of air is made up of cooler and heavier air. A low-pressure mass of air is made up of lighter and warmer air. Pressure differences cause air to move from high-pressure areas to low-pressure areas, resulting in wind. The wind speed can greatly affect the pollutant concentration in a local area. The higher the wind speed, the lower the pollutant concentration. Wind can dilute pollutants and rapidly disperse them in the immediate vicinity.

Atmospheric stability is the vertical motion of the atmosphere. Unstable atmospheric conditions can result in vertical mixing. Typically, the air near the surface of the earth is warmer in the day time due to the absorption of solar energy. The warmer and lighter air from the surface rises and mixes with the cooler and heavier air in the upper atmosphere causing unstable conditions in the atmosphere. This constant turnover results in dispersal of polluted air. A stable atmospheric condition usually occurs when the air close to the ground is warmer than the air above it, and it therefore rises. When a static layer of cooler air is formed close to the ground, the mixing depth is significantly restricted. This condition is called temperature inversion. During a temperature inversion, air pollutants released into the lower layer are trapped and can be removed only by strong horizontal winds.

The dispersion of pollutants is also influenced by the amount of turbulence in the atmosphere near the source.

Turbulence could be due to the horizontal and vertical motion of the atmosphere. Some other factors that affect the concentration of air pollutants include: humidity, solar radiation and precipitation. Solar radiation contributes to the formation of ozone and acts to create secondary pollutants in the air. Humidity and precipitation can aid in the formation of substances responsible for acid rain.

The techniques used to disperse air pollutants are smokestacks that are tall industrial chimneys. They release pollutants high above the earth's surface so that emitted pollutants can sufficiently disperse in the atmosphere before reaching ground levels. The greater the stack height, the better the pollutants will be dispersed and diluted before affecting nearby ecosystems and populations.

Last but not the least, pollutant dispersion and transport can be negatively affected by climate and geographical factors. Classic examples of how climate can affect pollutant dispersion are the temperature inversions. This can prevent the rise and dispersal of pollutants from the lower atmosphere and cause localized air pollution. Examples include the air pollution disasters that took place in London and Donora, Pennsylvania.

Sources and Effects of Pollutants - Primary and Secondary Pollutants

The industrial revolution around the 18th century has brought about a drastic change in the chemical composition of our atmosphere primarily due to the use of fossil fuels for energy production and transportation. Air pollution, is not a localized environmental problem but is a problem affecting the whole globe. It causes complex health problems and individual pollutant effects vary from each other. The Indian Air (Prevention and Control of Pollution) Act, 1981 defines air pollution as: Any solid, liquid or gaseous substance (including noise) present in the atmosphere in such concentrations that may or tend to be injurious to human beings or other living beings or other living creatures or plants or property or enjoyment. Air pollutants are airborne particles that occur in concentrations high enough to threaten the health of people, plants and animals, buildings and historical monuments, or that causes toxicity to a given environment. Air pollutants being in the air medium are inhaled, hence infecting the lungs and the respiratory system. Subsequently, they are taken up by blood and pumped to all parts of the body. Air pollutants can also be deposited on soil, plants, and in water because all the above mediums are interconnected and further contribute to human health risks.

Some Classic Case Studies: Air Pollution

1. The London smog of 1952: A classic air pollution case study

During the cold month of December, the people of London burnt large amounts of coal for heating their homes, which resulted in a huge amount of smoke, soot, ash, sulphur oxides etc. generated from the chimneys. On December 5th 1952, the temperature went down to -1°C and the humidity at 80%. There was a thick fog and the air near the ground was moist. Usually, the air close to the ground is warmer than the air above it, and therefore rises but in this particular event it resulted in the formation of a static layer of cooler air close to the ground, which is an atmospheric phenomenon called temperature inversion. The accumulation of smoke close to the ground was so much that the sunlight was totally cut off and the air stayed cool and static. The term smog means fog that has soot in it. Thousands of tons of black soot, sticky particles of tar and gaseous sulphur dioxide trapped in the fog gave it a yellow-black colour. The water from the fog condensed around the soot particles. The sulphur dioxide reacted inside the foggy, sooty droplets forming sulphuric acid producing intense acid rain. Average smoke measurements taken at the National

Gallery in London between the 4th and 8th December 1952 showed that the PM_{10} concentration was 56 times the level normally experienced and the sulphur dioxide level increased by 7 times (700 ppb). Around 4000 people died due to the smog causing pneumonia, bronchitis, tuberculosis, heart failure, asphyxiation, chest pains, inflammation of the lungs, damage to respiratory cells, permanent lung damage, respiratory ailments, susceptibility to cancers etc. The effects on vegetation were profound due to the resulting acidic rain. Following this disastrous event, London formulated the Clean Air Act, 1956 and all the traditional coal fires were converted to heaters fueled by gas, steam, hot water, oil, smokeless coal and electricity.

2. The terrible Bhopal gas Tragedy of India, 1984

Another classic case study involving air pollution is one of India's terrible industrial disasters, which occurred on 3rd December 1984 in Bhopal, Madhya Pradesh. A Union Carbide Company used methyl isocyanate in manufacturing Carbaryl (carbamate) pesticides. On the fatal day none of the safety devices worked and there was a functional failure of the vent scrubber outlet. There was a violent chemical reaction in the tank and the pressure built up. This popped the safety valve open. The safety valve remained open for more than two hours releasing 50,000 lbs of gaseous and liquid MIC along with HCN and $COCl_2$. As a result over 30 - 40 tons of methyl isocyanate (MIC) and 35 – 40 kg phosgene were emitted into the atmosphere, which could be seen as thick foggy dense gas clouds. The gas spread over the entire region and city, air being the medium. Approximately 5000 people died in the disaster. More than 2,50,000 people were exposed to the gas. MIC afflicts the lungs, eyes, skin, respiratory system, gastrointestinal system etc. Also, it causes reproductive, teratogenic and mutagenic effects. The clean up of the tragedy costed approximately 570 million

U.S. dollars compared to the cost of the safety devices, which would have costed only 1 million U.S. dollars if it had been installed before the tragedy. The after effects of the Bhopal gas tragedy are seen even today with people and children having genetic disorders.

3. Taj Mahal, Agra, India: A Symbol of Love affected by Air Pollutants

Taj Mahal means Crown Palace. Emperor Shah Jahan constructed this historic monument in the memory of his dear wife Mumtaz Mahal at Agra, India as an expression of love for his wife. It was constructed over a period of twenty-two years, employing twenty thousand workers and was completed in the year 1648. The innumerable industries in and around Agra released a large amount of sulphur dioxides, suspended particulate matter, smoke, soot etc causing tanning, blackening, and yellowing of the marble stone. The oxides reacting with rain water resulted in acid precipitation and in return caused the deterioration and corrosion of the Taj Mahal. In 1984, M.C. Mehta, an environmental lawyer filed a public interest litigation regarding the effect of pollutants on the historic monument. Following this, the Supreme Court of India ordered 292 industries in the Taj area to be shut down and or adopt cleaner control technologies.

Most of the populated metropolitan cities like Mumbai, Chennai, New Delhi, Kolkatta are heavily polluted. During peak office hours the air is hazy, smoky and suffocating and is a serious problem. The Air Pollution Control Act, 1981, was passed to control and abate air pollution. Following the Bhopal gas tragedy, the Environment Protection Act, 1986 was passed. The Central Pollution Control Board passed its national ambient air quality-monitoring program, 1985 and is being monitored by National Ambient Air Quality Monitoring Programme (NAAQM). These acts have focused on improving the quality of vehicular fuels, awareness among the general public, using renewable sources of energy and in the control of pollution.

Sources of Air Pollution

Automobile exhausts, industries, growing population, modernization, are a few major factors responsible for air pollution. They consist of inorganic, organic, gaseous pollutants, SPM (suspended particulate matter) such as dust, fumes, mist, and smoke. Anthropogenic air pollution sources are industries, energy generation, power plants, transportation etc. Since these pollutants are generally concentrated in and around urban areas, the outdoor urban pollution levels are higher than that in the rural areas lacking these industries. Fires-including forest fires, oil well fires, burning of leaves, large-scale burning of agricultural waste- are another major source of air pollution and can lead to severe problems if the smoke is inhaled for a period of time.

The two important types of air pollutants based on the sources are: stationary and mobile sources. The stationary sources as the name indicates, are stationary which includes pollution from point sources (specific point, i.e. chimneys, power plants), area sources (specific area, i.e. an industrialized area) and fugitive sources (open areas exposed to wind). The mobile sources as the name indicates, are mobile which includes emissions from aircrafts, ships, automobile emissions etc.

Table 3.3 : Some atmospheric pollutants and their natural sources

Natural Sources	Pollutants
Wind	Suspended particles
Volcanic eruptions	Dust, ash, gases like SO_2, CO_2
Forest fires	Smoke, ash, unburned hydrocarbons, CO_2
Vegetation	VOC's, pollen grains
Ocean Waves	Salt particles

Table 3.4 : Some atmospheric pollutants and their anthropogenic sources	
Anthroprogenic Sources	**Pollutants**
Industrial (Paper mills, power plants, refineries)	Particulate matter, SOx, NOx, ash
Domestic/ Residential (Transportation, home furnaces, Fireplaces)	CO, NOx, VOC's, particulate matter, Open burning of refuse
Coal mining/ Oil and gas production	Suspended particulate matter, oxides of nitrogen and Sulphur, volatile organic compounds
Power generation/ Vehicular sources	Suspended particulate matter, sulphur dioxide, Nitrogen oxides, carbon monoxide, VOC's, lead
Community services (Incinerators)	Suspended particulate matter, sulphur dioxide, Lead, VOC's
Agriculture/ Horticulture	Particulate matter, carbon monoxide, volatile organic compounds (VOC's)

Classification of Air Pollutants

Recently scientists discovered pesticides in Antarctica, where pesticides have never been used!!! This shows how pollutants are transported from one place to another. The source of pollution could be in one country but the impact may be felt or seen elsewhere. Volcanic activity is the most important natural source of air pollution releasing large amounts of ash and toxic fumes into the atmosphere. The volcanic eruptions of Krakatoa in Indonesia, Mt. St. Helens in Washington, USA and Katmai in Alaska, USA, have been related to measurable climatic changes.

Air pollutants mainly include gases, aerosols, and particulate materials. They can also be classified as primary and secondary pollutants. Primary pollutants are directly emitted into the air and include carbon monoxide, sulfur and nitrogen oxides, hydrocarbons, particulates etc. They are direct products of combustion and evaporation. When the primary pollutants undergo further chemical and photochemical reactions with different atmospheric compounds, secondary pollutants are formed like ozone and reactive organic compounds.

Table 3.5 Classification of air pollutants	
Classification of pollutants	**Examples**
Primary pollutants	Carbon monoxide, Sulphur oxides, Nitrogen oxides, Hydrocarbons, Particulates
Secondary pollutants	Ozone, peroxyacetyl nitrates, volatile organic compounds

A. Primary pollutants

1. Carbon monoxide (CO): Sources, Effects and Control

Sources: Around 90% of CO comes from natural sources while 10% of the CO in our atmosphere is contributed by forest fires, automobile emissions, incomplete burning of organic compounds, and incomplete combustion of fossil fuels. The other sources include the wood industry and heating from residences. The air in metropolitan cities air may contain average amount of 100 ppm or more of carbon monoxide during peak traffic hours due to fuel combustion. Catalytic converters are therefore used in modern motor vehicles that reduce carbon monoxide emissions. CO is a colourless, odourless gas and is toxic to human beings and animals. Carbon monoxide has a lifetime of four months in the atmosphere, and is 96.5% as heavy as air.

Fate of CO in the atmosphere: It undergoes reaction with the hydroxyl radical to form carbon dioxide. The atmospheric reactions are very slow and do not account for their substantial removal. Some microorganisms are able to remove CO rapidly. The permissible limit for CO in the ambient air is 32 ppm.

Effects: CO results in nausea, dizziness, suffocation, vomiting and in acute cases death as it has a natural affinity for hemoglobin (around 250 times more than oxygen). They together form carboxyhaemoglobin, which disables the transport of oxygen to the different parts of our body. The binding with other haeme proteins causes changes in the function of the affected organs such as the brain and the cardiovascular system, and also can affect the developing foetus. It can impair concentration, reflexes, impaired visual perception, manual dexterity and learning disability.

Control: Replace old polluting vehicles, promote the use of catalytic convertors in vehicles, and adopt cleaner fuel technologies. A brief account of the catalytic converter is described below.

Catalytic Converters

With increasing fossil fuel consumption due to rise in the number of vehicles and stringent emission standards, modern vehicles are equipped with three-way catalytic converters as it helps in three regulated emissions-of carbon monoxide, VOC's and NOx oxides. The catalytic converter is fitted onto the exhaust pipes of the cars and works only on unleaded fuel. This technology was made mandatory in the newly registered cars in the four main metros in India mainly to reduce air pollution. The converter uses two different types of catalysts; a reduction catalyst and an oxidation catalyst consisting of ceramic structures coated with a metal catalyst, usually platinum, rhodium and/or palladium. The first stage uses platinum and rhodium to reduce the NOx emissions. When an NO or NO_2 molecule contacts the catalyst, it rips the nitrogen atom out of the molecule, freeing the oxygen in the form of O_2. The nitrogen atoms bond with other nitrogen atoms that are also stuck to the catalyst, forming N_2. The second stage of the catalytic converter is the oxidation catalyst reduces the unburnt hydrocarbons and carbon monoxide by oxidizing them over a platinum and palladium catalyst. The third stage is a control system that monitors the exhaust stream to control the fuel injection system. There is an oxygen sensor, which tells the engine computer, how much oxygen is in the exhaust and the engine computer can increase or decrease the amount of oxygen in the exhaust by adjusting the air-to-fuel ratio.

2. Sulphur oxides: Sources, Effects and Control

Sources: Fossil fuel burning, industrial power plants, petroleum refineries, etc are the major sources contributing to SOx emissions. It is produced from sulphur containing materials like, coal, oil and metal ores used in combustion and refining processes, fly ash from coal, iron and steel industries, solid waste disposal etc. SO_2 can dissolve in water vapour in the atmosphere forming acids, interacting with other gases and particles in the air forming sulphates harmful to the environment. SO_2 is oxidized to SO_3 and it exists in the atmosphere when moisture is present. Approximately 1/3rd of the sulphur emitted into the atmosphere is from anthropogenic activities and around 2/3rd is from natural sources which include H_2S and SO_x

Effects: SO_2 causes health problems like lung damage, bronchitis and several respiratory disorders, plant disorders and damage to barley, alfalfa etc, corrosion of metals, marbles, leather, paints etc. SO_2 in the air leads to diseases of the lung, emphysema and other lung disorders such as wheezing and shortness of breath. When it interacts with ammonia it causes respirable particulate matter which has harmful effects on human health, environment, and contribute to visibility impairment and a hazy atmosphere. It can cause damage to vegetation, buildings and acidify the aquatic and terrestrial ecosystems. It also causes damage to fabrics, leather and paints. They are transformed into acidic sulphate particles that are deposited on aquatic and terrestrial ecosystems.

Control Measures: Scrubbing technologies have been used since 1970's to remove or reduce SO_2 from power plant emissions, popularly called as Flue Gas Desulphurization or FGD. Scrubbing is normally carried out with a buffered solution to neutralize the SO_2 and can remove both particulate and SOx. This is achieved by using sodium (sodium hydroxide, sodium carbonate, and sodium bicarbonate), lime or seawater. The waste stream generated from the scrubber is treated for removal of solids as well as biological oxidation demand before being discharged into the municipal sewer system. All the present technologies involve exposing combustion gases to a substance that will absorb most of the SO_2. FGDs can be a wet technology where the flue gas is exposed to an aqueous solution and a dry technology where the flue gas is exposed to solid absorbents. The wet technologies are more efficient in removing SO_2 over the dry technologies. Industries could reduce the amount of coal usage. Planting more trees can help in absorbing the pollutants.

3. Nitrogen oxides (NO_x): Sources, Effects, and Control

Sources: The major nitrogen oxides are N_2O (Nitrous oxide), NO (Nitric oxide), NO_2 (Nitrogen dioxide), NH_3 (Ammonia), N_2O_3, N_2O_5. The major contributors are automobiles and fossil fuels burnt by power plants, natural bacterial action in soils, natural burning, forest fires, anthropogenic sources etc. They are formed from nitrogen containing fuel can dissolve in water vapour to form acids, forms nitrates and other products detrimental to the environment. Under high pressure and temperature conditions in an engine, nitrogen and oxygen atoms in the air react to form various nitrogen oxides, collectively known as NOx. They are precursors to the formation of ozone and form nitrous oxide, which is 300 times more potent than carbon dioxide as a greenhouse gas. Nitrous oxide (N_2O) makes up about 7.2 percent of the gases responsible for global warming. Nitrous oxide also comes from nitrogen-based fertilizers and manure from farms.

Fate of NO_x in the atmosphere: They themselves may not be detrimental but when they combine with hydrocarbons, oxygen etc in the photochemical reactions they contribute to ozone formations and smog.

Effects: NO_2 is a reddish brown gas, highly toxic and corrosive in nature. This gas can make children susceptible to respiratory diseases in the winters. NO_2 itself can cause adverse effects on respiratory systems of humans and animals, and damage to vegetation. It can dissolve in water vapour and form acids (HNO_3) that have adverse effects on the respiratory systems of humans and animals. HNO_3 nitric acid, can cause severe damage to vegetation, buildings and acidify the aquatic and terrestrial ecosystems. Nitrate can combine with other compounds in the atmosphere, for example, ammonia, and form respirable particulate matter ($PM_{2.5}$), which is detrimental to the health of humans.

Control: The NOx emissions can be controlled by primary measures or flue gas treatment technologies.

Primary measures in a power plant include the following :

- Burner optimization for minimizing the formation of NOx during combustion
- Air staging and flue gas recirculation to reduce the oxygen availability thereby helping to reduce NOx formation and complete combustion
- Fuel staging where the previously formed NOx is reduced to nitrogen and oxygen
- Low NOx burners

These are considered integral parts of a newly built power plant and effectively reduce their NOx emissions.

4. Hydrocarbons: Sources, Effects and Control

Sources: Hydrocarbons are the simplest organic compounds consisting of carbon and hydrogen. They can be straight chain, branched chain, or cyclic molecules. There are solid, liquid and gaseous hydrocarbons. They include natural gas or methane, propane, butane, terpenes, etc. Approximately 80% of the hydrocarbons emitted are from natural sources like swamps, cattle sheds, landfills, fields etc. The anthropogenic sources for hydrocarbons are from automobiles. Methane constitutes 18% of the total greenhouse gas emissions and remains for as long as 10 years in the atmosphere. Hydrocarbons can interact with sunlight forming photochemical smog.

a. Solid hydrocarbons

They include crude oil, paraffins, asphaltic material, tar etc. Melted tar, which is very important in road building material, creates pollution problems.

b. Liquid hydrocarbons

They include oil, grease etc. Oil spillages are common phenomena's in the sea. They have affected the sea birds and marine ecosystems. The Gulf war and Persian Gulf oil slicks are the major oil slicks known.

c. Gaseous hydrocarbons

They are classified into volatile and non-volatile hydrocarbons. They can also be classified as polycyclic aromatic hydrocarbons and $C_1 - C_{10}$ hydrocarbons. C_1 hydrocarbon is methane and $C_2 - C_{10}$ hydrocarbons are called non-methane hydrocarbons (NMHC).

➢Methane

It is a gaseous hydrocarbon released during bacterial fermentation processes. The natural sources include: forest fires, paddy fields, agricultural fields, fermentation products, etc. It is a greenhouse gas and plays an important role in global warming and climate change.

➢Polycyclic Aromatic Hydrocarbons (PAHs)

Incomplete burning of coal, oil and gas, waste, garbage, tobacco, volcanoes, forest fires, automobile exhaust, etc results in the formation of Polycyclic aromatic hydrocarbons (PAHs). They are also found in many food items like vegetable oils, fish, meat, beverages, smoked, grilled fish and meat. PAHs are normally found as a mixture containing two or more of these compounds, do not dissolve easily in water and stick to solid particles and settle down the bottoms of lakes and rivers. In the soil, PAHs stick tightly to particles and contaminate ground water. PAHs can break down by reacting with sunlight and other chemicals in the air, over a period of days to weeks and are responsible for depletion of the protective ozone. They are known to cause cancers, teratogenic (birth defects) effects etc.

➤**Volatile organic compounds (VOC)**

They include carbon-containing gases and vapors such as gasoline fumes, paint solvents and organic cleaning solutions. They enter the atmosphere by evaporation for example as vapour, and as molecules resulting from the incomplete burning of fuels or wastes. The more reactive VOC´s combine with nitrogen oxides (NOx) in the photochemical reactions in the atmosphere and result in the formation of ground-level ozone, a major contributor of smog. Many VOC are known to cause direct toxic effects on human beings, causing carcinogenesis and in some cases neurotoxicity. A number of individual VOC, for e.g. benzene, dichloromethane are extremely toxic.

Effects:

Hydrocarbon pollution due to oil tankers spillage etc can cause smothering of benthic organisms (clams, oysters, worms). Some hydrocarbons caused narcosis and anesthesia in lower animals. They are cellular/ sub cellular poisons, cause gonadal tumors in clams, inhibited reproduction in mussels, fish cancers induced by benzapyrene, benzanathrocene carcinogens, etc. The coating on the birds that help in insulation and hypothermia is lost; as a result the birds lose buoyancy and sink. They cause necrosis, epinasty (downward curvature of leaf), chlorosis (yellowing of leaf), abscission (dropping of leaves), damage to the cuticles and stomata in plants.

Control:

These systems involve the oxidation process, in which the VOC pollutants within the air stream are broken down from their original composition and reconstituted into new compounds. In the VOC thermal oxidation process, as the name indicates, enough heat and oxygen are added to the hydrocarbons to create the oxidation reactions. VOC Hydrocarbons consist of carbon and hydrogen; they are converted naturally into carbon dioxide, water vapour and heat energy. The heat energy is then recuperated into the system by use of a heat exchange device, while the clean air stream of carbon dioxide and water is discharged to the atmosphere. Carbon adsorption and incineration of hydrocarbons can be used to reduce emissions.

5. Particulate matter (PM)

They are airborne particles in solid or liquid form. They may be classified as primary or secondary particulate matter depending on the compounds and processes involved during its formation. Primary PM is emitted at the emission source in particle form, i.e. the smokestack of an electrical power plant. Secondary PM formation results from a series of chemical and physical reactions involving different precursor gases, such as sulphur and nitrogen oxides, and ammonia reacting to form sulphate, nitrate and ammonium particulate matter.

The size of PM particles largely determines the extent of environmental and health damage caused. Some terms used are described below.

- **Total particulate matter (TPM):** airborne particulate matter with an upper size limit of approximately 100 micrometer (µm) in aerodynamic equivalent diameter.

- **Particulate matter <10 microns (PM_{10}):** airborne particulate matter with a mass median diameter less than 10 μm.
- **Particulate matter <2.5 microns ($PM_{2.5}$):** airborne particulate matter with a mass median diameter less than 2.5 μm.

Respiratory diseases such as asthma, bronchitis and emphysema and to various forms of heart disease have been linked to PM. They can also have adverse effects on vegetation and structures, and contributes to visibility deterioration and regional haze.

Types of particulate matters

Particulate matters are finely divided solid particles, liquid droplets, or a combination of the above. Water that is combined with a contaminant, such as acid mist, is considered particulate matter.

➤ **Aerosols**	Solid or liquid particles smaller than 1um.
➤ **Condensation Aerosol**	Aerosol formed by condensations of vapor or reactions of gases.
➤ **Dispersion Aerosol**	Formed by grinding solids, dispersion of dust, and atomization of liquids.
➤ **Fog**	High level of water droplets.
➤ **Haze**	Decreased visibility due to presence of particles.
➤ **Mist**	Liquid particles smaller than 10 um.
➤ **Fumes**	Solid particles ranging from 0.1 – 1um
➤ **Smoke**	Particles resulting from incomplete combustion function ranging from 0.01 – 1 um.

- **Petrocoke** — Consists of 90% fixed carbon and remaining 10% volatile matter with sulphur and other toxic trace metals. It is released from oil refineries.
- **Fly-ash** — This consists of finely divided particles of ash, incompletely burned coal and the carbon content of fly-ash varies from 5 – 20%, size varies from 0.02m to over 300 m.
- **Dust** — Soil, coal dust, mineral, rock particles ranging from 1 – 200um.

Particulate matter enters the atmosphere from many different sources. Wind blown dust is the most obvious source of particulate matter. Wind blowing over the oceans picks up salt particles, and all combustion processes release both solid matter (soot, smoke, fly ash) and gases which react to form larger molecules that can be classified as aerosols or particulate matter.

The chemical compositions, size of the particles and physical properties of an aerosol governs its toxicity. Density and diameter determine the most important aspects of an aerosol's behavior in the atmosphere. Aerosol diameter is a descriptive term, because most particulates are non-spherical and irregularly shaped. Stoke's Law is the relationship that relates the settling rate to a particle's density and diameter. Stoke's Law tells us that, when all other things are constant, dense particles settle faster, larger particles settle faster, and a denser or more viscous air causes particles to settle slower. The application of the Stoke's law is employed in designing control equipments, settling chambers, etc in the removal of particulate matter.

Effects

Particulates affect health by transporting toxic substances into the respiratory tract. Toxic compounds (lead, cadmium, beryllium, PAHs, etc) can then be adsorbed into the body where they affect biological processes.

Control

The following technologies are commonly used to control particulate emissions:

- Electrostatic precipitators
- Gravitational settling chambers
- Wet collectors
- Centrifugal collectors
- Fabric filters
- Wet particulate scrubbers
- Mechanical collectors or cyclones
- High temperature, high-pressure particulate control devices

To give an idea of the control technologies, some of the commonly used methods in industries are briefly described.

Electrostatic precipitators

Industrial operations produce exhaust gases that contain dust, fly ash, fumes, fine particles like cadmium, sulfur, lead, coal tar, which are detrimental to human health and the environment. A widely used method of removing these contaminants is to use an electrostatic precipitator (Fig 3.3). They use electrical energy in the removal of particulate matter. They can be classified as low voltage (6000 – 12000 V, used in hospitals, commercial establishments) and high voltage (above 30000 V, used in refineries, industrial operations) units.

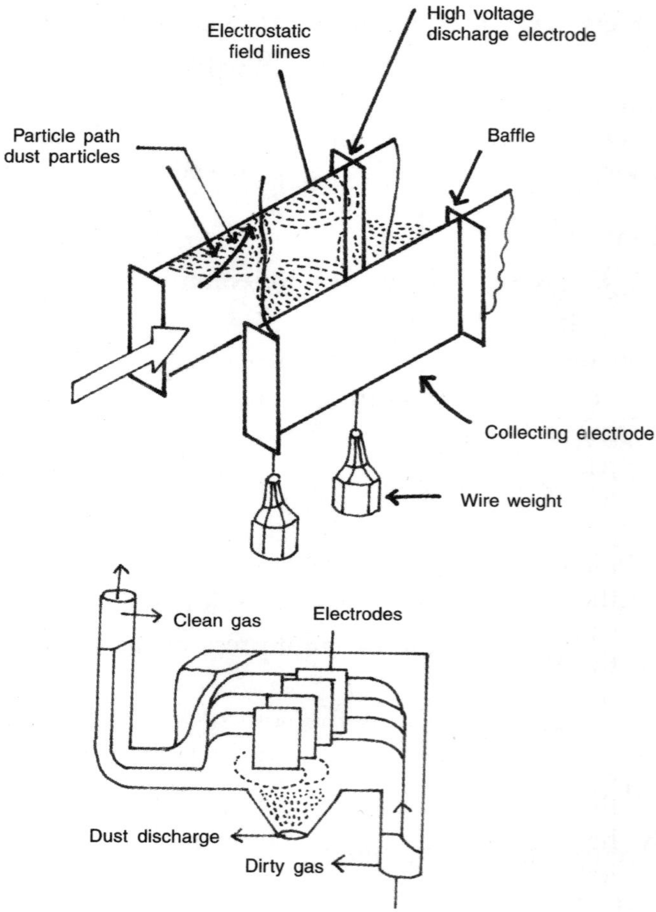

Fig. : 3.3 — Electrostatic precipitator

- **High voltage source**: A battery and solar cell produce direct current in which the current flows in one direction.
- **Discharge electrodes and collecting electrodes** – They are within but insulated electrically from the rest of the chamber and are charged with high direct current. There are plates, tubes and the inner walls, which serve as electrodes in particle collection.

- **Inlet and outlet for gas stream** – Facilitates entry of gas stream and removal of particulates.
- **Hopper**: To dispose of collected material.
- **Automated electronic cleaning system**
- **An outer enclosure / casing for the electrodes**

The gas stream is passed between the two electrodes across which a high potential difference is maintained. As a result a high ionizing field is formed. The particles caught in the flow of gas move through a collector chamber containing sets of electrodes in the form of parallel plates, bundles of tubes, or the collector's inner walls. The particles get electrically charged. The electrical charge ionizes (charges) the suspended particles, causing them to move toward the collecting electrodes. In a simple electrostatic precipitator, opposite high voltages are charged on two plates or grids. The positive grid charges the particles and the negative grid attracts (collects) them. If the material collected is dry, the collecting electrodes are tapped to loosen the particles, which fall into hoppers for collection and disposal.

They are very useful in the paper mills, cement factories and refineries. Their operation costs are low and cleaning the system is easy. The disadvantages include: explosion hazards and the production of toxic gases during gas ionization.

Some other important technologies used in air pollution control

- **Settling chambers:** It is used to remove high-density particulate material, for example mineral dust, grains etc. It works on the principle of gravitational force and settling velocities. The efficiency is quite low.
- **Dry Centrifugal collectors:** It uses the principle of centrifugal and centripetal forces. They can remove particulates effectively and are much better than the settling chambers.

- **Scrubbers:** Here, water is used to wet the particles usually in the form of water sprays. Water is also called the scrubbing medium. When a polluted gas enters the wet collectors, they go through an area containing water droplets. The dust settles upon the droplets and are collected and removed. The clean gas comes out thus helping in pollution control. They are useful in controlling acid mists, rock dusts, etc. They are very expensive but are highly efficient.
- **Fabric filters**: They are also known as baghouse filters. Here, the polluted gas stream passes through a felted fabric. The particles adhere to the fabric by electrostatic attraction, gravitational settling and diffusion. Fabric filters can be tubular; envelope shaped and is capable of removing particles as small as 0.1 um. They are used in cement industries, foundries and steel furnaces. They are efficient but high temperature gases (above 450°C) have to be cooled to lower temperatures (100 – 450°C).

6. Lead and effects

This is another important primary pollutant. It is a dull gray, odourless, tasteless heavy metal. In Indian gasoline approximately 0.15 g lead/ litre of gasoline is added as an anti knock agent or as an inexpensive way to prevent engine knock. The presence of lead in the atmosphere is primarily due to anthropogenic activities include processes such extraction / processing of metallic ores, incineration of solid waste and fuel combustion. The largest source of lead in the atmosphere is from leaded gasoline combustion, emissions from iron and steel production and lead smelters. Lead in drinking water is due to the presence of lead in pipes, solder, and fixtures. It is a pollutant of concern due to its persistence in the environment, potential for bioaccumulation, and its toxicity to humans and the environment. Lead is used in the manufacture of batteries, metallic products, paints, and ceramic. The most important sources of lead are due to the ingestion of lead contaminated foods and beverages, inhalation of airborne lead, etc. Lead is a highly toxic element, causing a variety of effects at low dose levels. Children are at higher risk as they put lead containing toys in their mouths.

Effects

Health effects of lead include intestinal cramps, peripheral nerve paralyses, anemia, severe fatigue, brain damage, kidney damage, gastrointestinal, blood pressure and disturbances in Vitamin D metabolism. Children show reduced growth and development. Reproductive effects with high lead exposure include decreased sperm count in men and spontaneous abortions in women.

B. Secondary pollutants:

Ozone and effects

Although ozone (O_3) is a desirable substance in the stratosphere, the same is a major environmental hazard at ground level. They are formed as result of chemical reactions between nitrogen oxides and volatile organic carbons with sunlight. Since sunlight is involved in the reactions, the products formed are collectively termed as photochemical oxidants. Ozone is a by-product of photochemical smog and reacts with hydrocarbons to form peroxynitrates that damage sensitive biological tissues in the eyes, nasal passages, throat and lungs. Excessive ozone levels in the troposphere have been responsible for the death of plants through reactions with chlorophyll. Ozone is formed naturally when oxygen molecules are photo chemically dissociated into oxygen atoms that can then react with a second oxygen molecule to make ozone. The presence of nitrogen oxides (NO, NO_2) leads to higher than normal background levels of ozone through photochemical reactions. The photochemical products such as ozone, peroxyacetyl nitrate, aldehydes and ketones have serious effects on human beings and animals (Fig 3.4). Ozone is not emitted directly but can be formed from reactions involving volatile organic compounds (VOC's), hydrocarbons and nitrogen oxides (NOx). Therefore, VOC's and NOx are called ozone precursors. G M B Dobson, in the 1920s, perfected an instrument to monitor total ozone. Hence ozone is measured in Dobson Units. 100 DU corresponds to a 1mm thick layer of ozone.

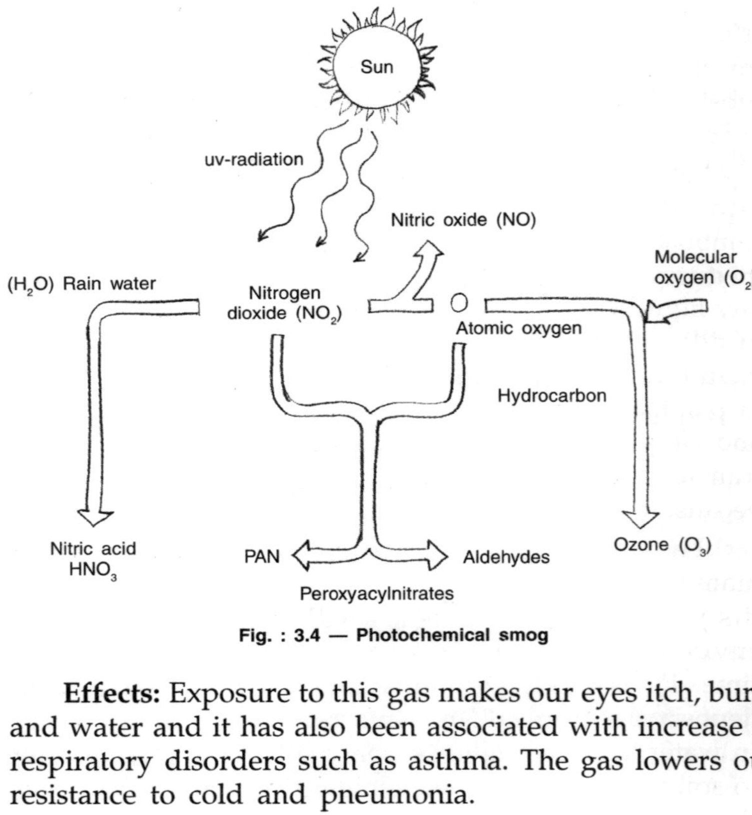

Fig. : 3.4 — Photochemical smog

Effects: Exposure to this gas makes our eyes itch, burn, and water and it has also been associated with increase in respiratory disorders such as asthma. The gas lowers our resistance to cold and pneumonia.

Health effects of some other important air pollutants

Persistent organic pollutants (POPs): Persistent organic pollutants (POPs) are semi-volatile chemicals such as PCBs, pesticides like DDT, chlordane and toxaphene, dioxins and furans. They enter the environment as a result of human activities and can bioaccumulate in living organisms, persist in the environment and produce long-term toxic effects.

When released into the environment, they are transported to great distances by evaporation and condensation popularly called the grasshopper effect. In the Arctic, the low evaporation rates trap POPs, thereby allowing them to enter the food chain.

Evidences indicate that POP´s cause significant adverse effects to human beings and wildlife. The Stockholm convention on the POP's was a global agreement to substantially reduce or eliminate emissions of around 12 toxic substances.

Polychlorinated biphenyls (PCB's): They are emitted into the environment through different industrial, combustion, energy and traffic related processes. They are mixtures of individual chlorinated compounds. They an be oily liquids, solids or vapours in the air. PCBs are used in hundreds of industrial and commercial applications including electrical and hydraulic equipment; as plasticizers in paints, plastics and rubber products; in pigments, dyes and carbonless copy paper, as coolants and lubricants in transformers, capacitors, and other electrical equipment because they do not burn easily and are good insulators.

They enter the air, water, and soil during their manufacture, accidental spills and leaks and their improper disposal. PCBs do not readily break down in the environment and may remain for very long periods of time. PCBs can travel long distances in the air and be deposited in areas far away from where they were released. In water, they stick to organic particles and bind strongly to soil particles. They accumulate in fishes reaching levels that are a thousand times higher than the original levels in water.

Health effects include acne, rashes, neurobehavioral and immunological changes. The main dietary sources include: consuming fish, meat, dairy products and drinking contaminated well water.

Radon: A radioactive gas that can accumulate inside the house, it originates from the rocks and soil under the house and its level is dominated by the outdoor air and also to some extent the other gases being emitted indoors. Exposure to this gas increases the risk of lung cancer.

Tobacco smoke: Tobacco smoke generates a lot of harmful chemicals and is a major cause of ill health, and known to cause cancer. It not only affects the smoker but affects the passive smokers (the person who is near / or in the vicinity of a smoker and is not a smoker) also. The effects include burning sensation in the eyes and nose, throat irritation, cancer, bronchitis, severe asthma, and a decrease in lung function.

Biological pollutants: They are mostly allergens like pollen grains, fungal spores, plant materials, animal fur etc. that can cause asthma, hay fever, dermatitis and other allergic diseases.

Some Issues Relating to Air Pollution

1. Greenhouse effect

Climate change refers to the changes in the long-term climate or changes in the average weather for a particular location. It can result from natural processes, such as the change in the sun's strength, and also from human activities, through the build-up of greenhouse gases. Environmentalists are concerned with the concentrations of the greenhouse gases, which act as a blanket and trap the heat that comes from the earth (Fig 3.5). This blanket effect is natural and a very important part of life on earth. However, the atmospheric concentrations of these gases have increased significantly since pre-industrial times largely because of fossil fuel usage, decrease in forest cover, etc resulting in climate change.

The earth receives energy from the sun, which warms the earth's surface. As this energy passes through the atmosphere, a certain percentage (about 30%) gets scattered. Some part of this energy is reflected back into the atmosphere from the land and ocean surface. The rest (70%) remains to heat the earth. In order to establish a balance,

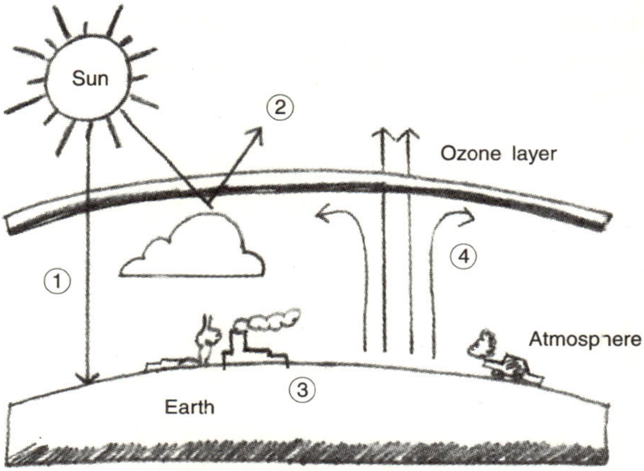

Fig. : 3.5 — The Greenhouse effect

the earth must radiate some energy back into the atmosphere. As the earth is much cooler than the sun, it does not emit energy as visible light but emits through infrared or thermal radiation. However, certain gases in the atmosphere form a sort of blanket around the earth and absorb some of this energy emitted back into the atmosphere. Without this blanket effect, our earth would be around 30°C colder than it normally is. These gases like carbon dioxide, methane, and nitrous oxide, along with water vapour, comprise less than one per cent of the atmosphere. They are called 'greenhouse gases' as the working principle is same as that which occurs in a greenhouse. The greenhouse gas layer functions like a green glass house, where the glass allows sunlight to come in but prevent the heat from going out. Glass easily transmits short wave length solar energy into the greenhouse and is almost opaque to the longer wavelengths radiated by the interiors of the greenhouse. This trapping of the radiation is responsible for the increased temperatures inside the greenhouse. This phenomenon is known as the green house effect and this process results in warming the air in the lower layers leading to global warming. This gas blanket has been in place ever since the creation of the earth. Since the industrial revolution human activities have been releasing more and more of these greenhouse gases into the atmosphere. This leads to the blanket becoming thicker and upsets the 'natural greenhouse effect'. The resulting enhanced greenhouse effect is more commonly referred to as global warming or climate change. The most important green house gases are CO_2, methane, Chlorofluorocarbons and nitrous oxide. Scientists predict 1.5 - 4.5°C increase in temperatures by 2100. This may result in glaciers melting, sea levels rising, and redistribution of dry and wet regions.

During the pre-industrial era, the amount of carbon dioxide remained constant but now it has increased from 250 ppm (1900) to approximately 350 ppm due to industrialization and urbanization. Economic activity, technology and population explosion are driving factors in enhancing global warming. Atmospheric CO_2 levels have been increasing at an alarming rate since 1970. 100 years back the CO_2 level was just 275 ppm, now it is approximately 350-370 ppm. It is expected to reach 450 ppm by the year 2040. This will result in an enormous increase in the earth's temperature. The Kyoto protocol formulated in December 1997 in Kyoto, Japan, focused on reducing carbon emissions by 50% or to the pre 1990 levels.

Table 3.6 : The Major greenhouse gases, their sources and atmospheric concentration

Major Green- -house Gases	Atmospheric concentration (ppm)	Sources
Carbon dioxide	351	Fossil fuel combustion, deforestation
Nitrous oxide	0.31	Fertilizers, fossil fuels, biomass burning
Methane	1.675	Swamps, marshes, paddy fields, wetlands, livestock, natural gas leaks, fermentation from

Contd...

Contd...

Major Green- -house Gases	Atmospheric concentration (ppm)	Sources
		enteric processes of animals, biomass burning
Chlorofluorocarbons	0.00225	Aerosols, plastic foams, industrial solvents, air conditioning, refrigeration

Environmental impacts of greenhouse gases and global warming

A rise in temperature would lead to the following consequences:

- Rapid loss of biodiversity and a decline in natural resources.
- There could be a change in the rainfall and monsoon patterns.
- Melting of ice caps and glaciers.
- Increasing temperatures would cause thermal expansion of the oceans and the sea level will rise. This will result in floods in the productive deltaic regions. Sea level rise can result in the disappearance of many islands in the Caribbean, Pacific and the Indian oceans.
- Climate change can cause disastrous effects on the ecosystems.
- Disruption of marine ecosystems and flooding of the coastal wetlands.
- It could result in frequent storms in many regions where storms were never experienced.
- Floods can result in the spread of infectious diseases.
- Groundwater can become contaminated.

2. Depletion of ozone layer

Ozone is a triatomic form of oxygen and is a strong oxidant. It derives its name from the Greek word *ozein* meaning smell as it has a pungent smell in a concentrated form. About 90 % of the ozone in the atmosphere is present in the stratosphere. This protective ozone is formed as a product of a photochemical equilibrium. Ozone is being continuously formed in the stratosphere by the absorption of short wavelength ultraviolet radiation and is also removed by chemical reactions converting it back to molecular oxygen. Ozone formation in the stratosphere is described by the following reactions:

$$O_2 + h\nu \rightarrow O + O$$

$$O + O_2 + M \rightarrow O_3 + M$$

Where, M represents a third body (N_2 or O_2) necessary to carry away the energy released in the reaction and hv is the ultraviolet radiation in this reaction having wavelength less than 242 nm. Opposing the above formation process is ozone removal by photo dissociation:

$$O_3 + h\nu \rightarrow O_2 + O$$

Wherein ozone absorbs UV-radiation in the 200 - 320 nm wavelength regions. The above combination of reactions form a long chain in which oxygen atoms are constantly being shuttled back and forth between various molecular forms. The net effect of the above reactions is the creation of a layer of ozone in the stratosphere. The balance between creation and removal is affected by increasing stratospheric concentrations of chlorine, nitrogen and bromine, which act as catalysts speeding up the removal process. The most prominent ozone destructive gases are the CFCs that are very stable compounds, relatively unaffected by the usual pollutant removal process in the troposphere. CFC molecules can be broken by ultraviolet radiation freeing the chlorine that is available to destroy ozone. For example, the reaction involving CFC-12 is as follows:

$$CCl_2F_2 + h\nu \rightarrow Cl + CClF_2$$

Ozone can be removed from the stratosphere by catalytic reactions involving chlorine, nitrogen, hydrogen oxides or bromines. It is the enhancement of these reactions by anthropogenic activities that are of concern. Catalytic destruction of ozone can be explained with the help of the following reactions:

$$\begin{array}{c} X + O_3 \rightarrow XO + O_2 \\ \underline{XO + O \rightarrow X + O_2} \\ O + O_3 \rightarrow O_2 + O_2 \end{array}$$

Where X may be Cl, Br, OH or NO. The net result of this pair of reactions is the destruction of ozone molecule. Stratospheric ozone depletion refers to the destruction of the earth's ozone layer and thereby reducing its ability to shield ultraviolet (UV) radiation from the Earth's surface.

Ultraviolet radiations are of three kinds UV-A, UV-B and UV-C. Of these UV-B is known to be most damaging to biological systems. The ozone layer present in the stratosphere filters out the harmful UV-B radiations from the Sun, thereby protecting life on our earth. This UV-B has disastrous effects on humans, animals, vegetation and the environment by causing DNA damage resulting in Humans and Animals - mutations, eye cataracts and cancers, Plants – reduction of yield in crops like sorghum, rice, cotton, wheat, soybeans etc, Materials – degradation of paints and plastics.

Pollutants, such as chlorofluorocarbons (CFCs), released from refrigeration, air conditioning systems, solvents, plastic foams, aerosols, propellants destroy this equilibrium and the ozone (O_3) present in the stratosphere. The result is the thinning of the ozone layer, particularly at the poles. In addition to reducing ozone concentrations, CFCs are themselves potent greenhouse gases and trap an enormous amount of heat. In addition the supersonic aircrafts flying at stratospheric heights can cause major disturbances in the ozone layer. In spite of the slow vertical mixing of pollutants some of the pollutants like CFCs enter the stratosphere and remain there for many years until they are converted to other products. They stay for about

60 -100 years in the stratosphere. This results in depletion of the layer. The Montreal Protocol on Substances that Deplete the Ozone Layer (1987) listed the substances that cause depletion of the ozone and called for about 50% reduction of CFCs by the year 2000.

3. Acid rain

Another common effect of air pollution is acid rain. When rainwater goes below a pH of 5.6, it is called acid rain as normal rainwater has a pH of 5.6 due to the dissolution of CO_2 in rainwater. The phenomenon occurs when sulphur dioxide and nitrogen oxides from the burning of fossil fuels such as, petrol, diesel, and coal combine with water vapour in the atmosphere and fall as acidic rain, snow or fog. These gases can also be emitted from natural sources like volcanoes. Acid rain causes extensive damage to water, forest, soil resources and even human health. They can be wet depositions, where the acidic substances are dissolved in rain, dew, fog, snow and dry depositions, where acidic particulate matter in dry form can be deposited on vegetation, monuments etc. Acid rain mobilizes heavy metals like cadmium, mercury, calcium etc in soils, rocks, sediments that are slowly leached and enter the surface water bodies. Many lakes and streams have been contaminated and this has led to the disappearance of some species of fish in Europe, USA and Canada as also extensive damage to forests and other forms of life. It is said that it can corrode buildings, monuments and be hazardous to human health. Because the contaminants are carried long distances, the sources of acid rain are difficult to pinpoint and hence difficult to control. For example, the acid rain that may have damaged some forest in Canada could have originated in the industrial areas of USA.

Standards

Standards are set to control or reduce pollutant's levels to levels that are safe and healthy. The Smoke Nuisance act,

1920 was the first known pollution act in India set by the Bombay Government. The Central Air (Prevention and Control of Pollution) Act 1981 was developed for strict monitoring of the air pollutants. The air quality standards vary from country to country. The following steps were taken by the Indian Government to control air pollution:

- Establishment of Ambient Air Quality Monitoring throughout India
- Notification of Ambient Air Quality Standards under Environment (Protection) Act.
- Improving fuel quality by eliminating lead from gasoline, reducing diesel sulphur, reducing gasoline benzene, etc.
- Introduction of alternate fuelled vehicles like CNG/LPG.
- Improvement of public transport system.
- Public awareness and campaigns.

National Air Quality Monitoring Programme (NAMP)

The Central Pollution Control Board, India has executed a nation-wide programme for air quality monitoring known as National Air Quality Monitoring Programme (NAMP). This network consists of 308 operating stations that include: 115 cities/towns in 25 States and 4 Union Territories of the country. The objectives are to determine the status and trends of ambient air quality; to ascertain whether the prescribed ambient air quality standards are violated, to obtain a sound understanding necessary for developing preventive measures. The programme also regularly monitors four important air pollutants i.e. SO_2, NO_2, Suspended Particulate Matter (SPM) and Respirable Suspended Particulate Matter (RSPM/PM_{10}), at all the locations. The meteorological parameters, such as wind speed and direction, relative humidity and temperature are also monitored along with air quality. The monitoring is being carried out with the help of Central and State

Pollution Control Boards, National Environmental Engineering Research Institute (NEERI), Nagpur.

Table 3.7 : National ambient air quality standards (CPCB)

Pollutants	Time-weighted average	Concentration in ambient air		
		Industrial Areas	Residential, Rural & other Areas	Sensitive Areas
Sulphur Dioxide (SO_2)	Annual Average* 24 hours**	80 μg/m³ 120 μg/m³	60 μg/m³ 80 μg/m³	15 μg/m³ 30 μg/m³
Oxides of Nitrogen as (NOx)	Annual Average* 24 hours**	80 μg/m³ 120 μg/m³	60 μg/m³ 80 μg/m³	15 μg/m³ 30 μg/m³
Suspended Particulate Matter (SPM)	Annual Average* 24 hours**	360 μg/m³ 500 μg/m³ 120 μg/m³	140 μg/m³ 200 μg/m³ 60 μg/m³	70 μg/m³ 100 μg/m³ 50 μg/m³
Respirable Particulate Matter (RPM) (size less than 10 microns)	Annual Average* 24 hours**	150 μg/m³	100 μg/m³	75 μg/m³
Lead (Pb)	Annual Average* 24 hours**	1.0 μg/m³ 1.5 μg/m³	0.75 μg/m³ 1.00 μg/m³	0.50 μg/m³ 0.75 μg/m³
Carbon Monoxide (CO)	8 hours** 1 hour	5.0 mg/m³ 10.0 mg/m³	2.0 mg/m³ 4.0 mg/m³	1.0 mg/m³ 2.0 mg/m³

* Annual Arithmetic mean of minimum 104 measurements in a year taken twice a week 24 hourly at uniform interval.

** 24 hourly/ 8 hourly values should be met 98% of the time in a year. However, 2% of the time, it may exceed but not on two consecutive days.

Some Measures to Reduce Air Pollution

- Walk to the nearby markets whenever possible
- Use bicycles for short distances
- Develop efficient public transport systems like trams, trains, and buses and reduce dependence on private cars
- Reduce the use of aerosols in the household
- Planting more trees in the garden and surroundings. *Money plant, Coleus blumeri, Phaseolus vulgaris* and *Ficus* absorb carbon monoxide. *Pinus* species and *Juniperus* metabolize nitrogen oxide emissions
- Using unleaded petrol
- Promote energy efficiency and conservation, cutting down on wasteful energy consuming activities, using safer and cheaper alternative energy sources (Biogas, gasohol, solar power, hydrogen gas etc)
- Industries should use electrostatic precipitators, fabric filters, scrubbers, or other technologies to remove particulate matter from smoke stacks
- Zoning – industries should be away from residential areas, schools and hospitals. For example the city of Bangalore is well planned with proper zoning of the industries (i.e. light, medium and heavy industries)
- Avoid unnecessary burning activities, i.e. burning plastics, etc
- Avoid using plastic carry bags, use jute or coir bags for shopping
- Buying cars fitted with catalytic converters
- Using CFC free refrigerators and air conditioners
- Avoid the use and burning of polystyrene cups
- Do not use CFC containing deodorants, perfumes and sprays
- Make a compost pit in your backyard

Some Interesting Facts

- Scientists have found that citric acid could make catalytic converters used in cars as good as new as it removed 82% phosphorus and 90% sulphur.
- Particulate pollution such as soot is 10 times more in India as compared to global average. In India 42 % of soot comes from cooking, 25 % from fossil fuels and 33 % accounts for open burning.
- An average adult is assumed to breathe approximately ten cubic metres of air per day
- Pacific Island Nations at a UN conference held recently at Mauritius plans to popularize coconut oil as a fuel. Coconut oil does not pollute, has a nice smell and costs less.
- A new study by the NASA (National Aeronautics and Space Administration) has found out that the world heats up by 0.2°C every 10 years. This is forcing the plant and animal species to move towards the north and south poles respectively.
- Out of the 35-40 million tons of flyash generated annually by thermal power plants in India, only 2-3 percent is utilized productively.
- Neem and Pipal are the largest emitters of oxygen and purify the surrounding air.
- Many commercial PCB mixtures were known by the trade name Aroclor in the USA. Due to their toxicity and persistence in the environment, the Polychlorinated Biphenyis (PCBs) led Congress in 1976 in the USA enacted the Section 6(e) of the Toxic Substances Control Act (TSCA) that prohibited the manufacture, processing and distribution of PCBs. The manufacture of PCBs was stopped in the U.S. in 1977 because of evidence they build up in the environment and can cause harmful health effects.
- In 1974, Scotland experienced an acidic rain (sulphuric shower), which was approximately pH 2.4 more sour than vinegar.
- Many air purifiers and fresheners contain the pesticide para-dichlorobenzene.

UNIT 4

WATER POLLUTION AND CONTROL

Our planet has been called Planet Ocean, planet water or the blue planet because of the abundant water on its surface. Liquid water is a rare commodity in the cosmos and there is no liquid water on the sun or anywhere else in the solar system or interstellar space. Liquid water can be found only on a planet of the right mass and chemical composition and at the right distance from the sun. Water is absolutely fundamental to life. Without water, it is difficult to imagine any form of life that may exist.

A few decades ago nobody had imagined drinking water being sold in bottles. But over the years due to rapid industrialization our water reservoirs have become contaminated. Today, the bottled water segment is the fastest growing area of the beverage market as there is a great consumer demand for bottled water. Globally, the market was estimated in 1995 to be worth more than $14 billion annually in wholesale sales, and it has certainly grown since then. The public is concerned about tap water safety and quality, and view bottled water as a purer, safer option. As potable water has become scarce, its proper management is important.

Hydrosphere

The earth's surface is 510 million square km; 362 million is covered by water and 325 million square kilometers by open oceans. The total amount of water on the Earth is 1.5 billion square kilometers out of which 93 % is locked in the oceans. Water forms two important aquatic environments, the fresh water and marine respectively. Water is vital for all living beings, in the formation of soil etc. Water has different residence times in various reservoirs and the largest reservoir is the oceans. The residence time is the average time a substance stays within a specified region of space, such as a reservoir. For example, the residence time of water stored in deep groundwater, is approximately 10,000 years.

Table 4.1 : Approximate residence times of water found in various reservoirs

Reservoir – Source	Average Approximate Residence Time
Snow cover, Rivers	2 to 6 months
Glaciers	20 to 100 years
Lakes	50 to 100 years
Groundwater	10,000 years

Natural Water: Some Unique Properties

1. The hydrosphere includes the water resources on the earth – oceans, seas, streams, lakes, polar ice caps, glaciers, rivers, reservoirs and ground water. 2/3 rd of the earth's surface is covered with water and the rest is the habitable land. 97 % of the earth's water is marine water and rest (3 %) is fresh water of which majority of it is locked in the polar ice caps and only 0.003 % is available to us in the form of surface and ground waters. Water is the major constituent of all life forms. Most plants and animals contain more than 60 % water by volume. Without water, probably there would have been no life on this planet.

2. Water has a very simple atomic structure consisting of two hydrogen atoms bonded to one oxygen atom. The nature of the atomic structure causes its molecules to have unique electrochemical properties.
3. The hydrogen side of the water molecule has a slight positive charge and the other side of the molecule has a negative charge. This molecular polarity makes water a powerful solvent and is responsible for its strong surface tension.
4. When water makes a physical phase change its molecules arrange themselves in distinctly different patterns (i.e. in ice). When water freezes its volume expands and density decreases. Expansion of water at freezing temperatures facilitates ice to float on top of liquid water.
5. Water has a high specific heat. This property causes water to absorb large amounts of heat energy before it begins to get hot and it releases heat energy slowly when situations cause it to cool. This property helps in the moderation of the earth's climate and helps organisms regulate their body temperature more effectively.
6. Pure water has a neutral pH. As a result, pure water is neither acidic nor basic but changes its pH when substances are dissolved in it. Rain has a natural acidic pH of about 5.6 because it contains naturally derived carbon dioxide and sulphur dioxide.
7. Water conducts heat more easily than any liquid except mercury. That is why lakes and oceans essentially have a uniform vertical temperature profile.
8. Water exists as a liquid over temperatures from 0 - 100° Celsius. This range allows water to remain as a liquid in most places on the Earth.
9. It is a universal solvent and dissolves large number of different chemical compounds. This feature enables water to carry solvent nutrients in runoff, infiltration, groundwater flow, and living organisms.

10. Water has a high surface tension. This property helps in the formation of water droplets and waves, allowing plants to move water, dissolved nutrients from their roots to their leaves, and the movement of blood through tiny vessels in the bodies of animals.
11. Water is the only substance that exists in all three physical states of matter: solid, liquid, and gas. It helps in the redistribution of heat energy in the Earth's atmosphere. In terms of heat being transferred into the atmosphere, approximately 3/4's of this process is accomplished by the evaporation and condensation of water.
12. Water is a transparent medium and facilitates light penetration.
13. Hydrological cycle – water circulates continually between the lithosphere, atmosphere in the hydrological cycle.

Table 4.2 : Properties and significance of natural water

Properties of water	Significance
Excellent solvent property	Transport of nutrients, waste products
High dielectric constant	High solubility of ionic substances, ionization in solution
High surface tension	Controlling factor in physiology
Transparent	Colorless, light penetration in aquatic bodies
Maximum density at 4°C	Ice floats; vertical circulation restricted in stratified water bodies
High heat of evaporation	Important in the transfer of heat and water molecules between the atmosphere and bodies of water
High latent heat of fusion	Temperature stabilization at freezing point of water
High heat capacity	Stabilization of temperatures of organisms

Hydrologic Cycle

The hydrologic cycle or the water cycle describes the storage and movement of water between the hydrosphere, biosphere, atmosphere and lithosphere (Fig 4.1). Water can be stored in any one of the following reservoirs: atmosphere, oceans, lakes, rivers, soils, glaciers, snowfields, and groundwater. Water is continually cycled between its various reservoirs. This cycling occurs through the processes of evaporation, condensation, precipitation, deposition, runoff, infiltration, sublimation, transpiration, melting, and groundwater flow. Of this evaporated water, only 91% of it is returned to the ocean basins by way of precipitation. The remaining 9% is transported to areas over landmasses where climatic factors induce precipitation. The resulting imbalance between rates of evaporation and precipitation over land and ocean is corrected by runoff and groundwater flow to the oceans.

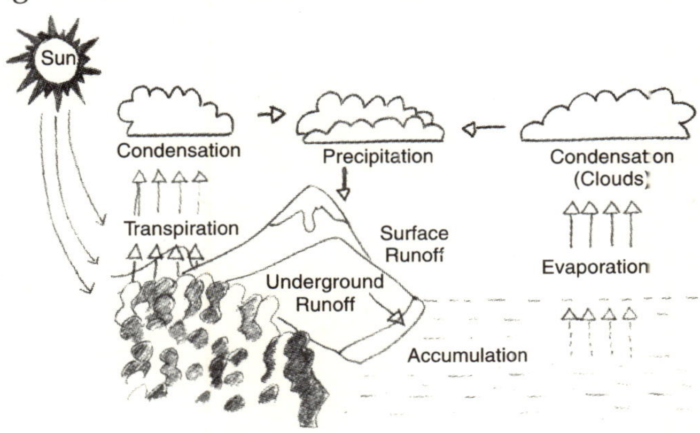

Fig. : 4.1 — Water Cycle

Water is constantly moving within and above the earth in a cycle called the hydrologic cycle. There are six major components of this cycle: evapotranspiration, condensation, precipitation, infiltration, percolation and runoff. The key concepts and components of the hydrologic cycle are as follows:

Evapotranspiration is the combined effect of two processes: evaporation and transpiration. It uses a larger portion of precipitation than the other processes associated with the hydrologic cycle. Evaporation is the process of returning moisture to the atmosphere. Water on any surface, especially the surfaces of mud holes, ponds, streams, rivers, lakes, and oceans, is warmed by the sun's heat until it reaches the point at which water turns into the vapour, or gaseous, form. The water vapour then rises into the atmosphere.

Transpiration is the process by which plants return moisture to the air. Plants take up water through their roots and then lose some of the water through pores in their leaves. As hot air passes over the surface of the leaves, the moisture absorbs the heat and evaporates into the air. Condensation involves cooling of water vapour until it becomes a liquid. When the dew point is reached, water vapour forms tiny visible water droplets to form clouds. As the droplets collide, they merge and form larger droplets and eventually precipitate.

Infiltration is the entry of water into the soil surface. It is the sole source of water to sustain the growth of vegetation and it helps sustain the ground water supply to wells, springs and streams. The rate of infiltration is influenced by the physical characteristics of the soil, plants, water content, soil temperature and rainfall.

Precipitation is moisture that falls from the atmosphere as rain, snow, sleet, or hail. It varies in amount, intensity, and form by season and geographic locations. These factors impact whether water will flow into streams or infiltrate into the ground. It is important for crop management as well as the engineering design of water control structures and flood control. Percolation is the downward movement of water through soil and rock. Percolation occurs beneath the root zone. Ground water percolates through the soil much as water fills a sponge, and moves from space to space along fractures in rock, through sand and gravel, or through channels in formations such as cavernous limestone. Runoff is the movement of water across the earth's surface towards stream channels, lakes, oceans, or depressions in the earth's surface. The characteristics that affect the rate of runoff include rainfall duration and intensity as well as the ground's slope, soil type and ground cover.

Water Pollution

Water-borne toxic chemicals pose the greatest threat to the safety of water supplies. Pollutants disturbing water bodies are domestic wastes or sewage, pesticides, industrial wastes, plastics, polythene, solid wastes, dead organic matter, acidic material, mine tailings, dyes etc. All the environmental pollutants in fresh water ultimately end up in the sea. Seventy-seven percent of the pollutants in the ocean come from the combined input of land runoff and atmospheric deposition. Only 12% of the ocean's pollutants come from shipping accidents while 10% are from ocean dumping. Off shore drilling accounts for the remaining 1%. This is significant when considering the potential impact on global fish production, a major source of protein for many countries. The sources of ground water contamination include: agricultural pesticide runoff, fertilizer runoff, industrial plant discharges, and drainage from both operational and defunct mines. The World Health Organization in Geneva, Switzerland establishes international water quality guidelines. Water quality standards include parameters in the three broad categories of biological, chemical and physical water quality. There are different standards written within these categories, because water may be designated for different uses. The defined uses for establishing water quality standards are: (1) domestic water (2) recreational uses (3) aquatic and wildlife uses and (4) agricultural use.

Water pollution means any undesirable characteristic or the degradation of water quality as indicated by physico-chemical and biological parameters. Used water from residential areas, industries and sewage is called wastewater. Domestic water includes that from baths, kitchen, dishwashers, washing machines and toilets. Small business establishments and cottage industries also contribute large amounts of wastewater to sewage collection systems. The turbid nature of sewage is due to suspended particles ranging from 100 to 350 mg/l. An important pollution parameter of the wastewater is the biochemical oxygen demand (BOD). The BOD is the amount of oxygen, which microorganisms consume to break down organic matter. Untreated sewage has a BOD of 100 mg/l to 300 mg/l.

Some Case Studies: Water Pollution

1. Deadly Arsenic contamination in Assam, West Bengal, and Bangladesh

The Assam Government in collaboration with the UNICEF collected 4200 drinking water samples in the region and sent them for laboratory analysis. The water was red in colour and contaminated with the heavy metal arsenic, which is a severe threat to the people in the region. Arsenic contamination was reported in Karimganj, Dhemaji and Dhubri districts of Assam. Arsenic when ingested in small quantities for 10-15 years manifests in white or black spots called melanosis. These spots get worsened into leprosy like lesions and skin ulcers. A very long exposure to arsenic causes lung cancers. Skin diseases are the most typical symptoms of chronic exposure to arsenic in drinking water. Symptoms include hyperkeratosis, skin cancer, renal, gastrointestinal, neurological, hematological, cardiovascular and respiratory problems. The whole belt of Assam, West Bengal and Bangladesh suffers from severe arsenic contamination of ground waters primarily due to geologic reasons and aggravated by the indiscriminate use of arsenic containing pesticides. The government has now taken steps to provide domestic water filters.

2. Textile dyes contaminating water in Coimbatore and Tirupur

Coimbatore and Tirupur are South India's major textile and cotton manufacturing centres, also known as the Manchester of the South. There are 700 odd dyeing and bleaching units in Tirupur generating approximately 50000 m^3/d of effluents. The discharge of effluents from bleaching and dyeing units has polluted the Noyyal River as well as the Ground water. The liquid effluents from South India Viscose Plant have affected the Bhavani River ecology. The industrial waste waters are discharged into the Cauvery River in Tiruchirapalli and Noyyal River (a tributary of River Cauvery) in Tirupur and Coimbatore. 850 polluting units in the region have failed in installing the common effluent treatment plant, which is mandatory. The effluents have polluted the waters of the Noyyal River and this contaminated water gets filled in the Ukkadam tank, thereby contaminating the area's groundwater. The chromium bearing effluents are toxic to the aquatic biota including fish. The organic constituents and sulphides are likely to affect aquatic ecosystems, through toxic effects and oxygen depletion. The TDS and the chromium in the effluent may affect the topsoil quality. As a remedial measure, the state government has directed the polluting units to invest in membrane technologies like reverse osmosis and keep the dissolved solids below 2000 parts per million in the effluents.

3. Marine pollution in Tamil Nadu: Oceans not spared

Industrial pollution has threatened the natural habitats of pearls in the pearl banks of Tuticorin coast in the Gulf of Mannar. It has affected fish and other organisms as far as 30 kms south of Tuticorin due to effluents released from chemical industries. Tannery wastes have caused the pollution of coastal waters from Chennai to Vedaranyam. The effect of diversity of phytoplankton ecology of mangrove estuaries of Tuticorin is greatly affected by industrial effluents. The Chennai coastal waters showed high levels of pesticides like DDT, lindane, endosulphan and heptachlor. The bioaccumulation of these pesticides in marine organisms could pose major health hazards.

4. The Holy Ganges is polluted

The river originates in the Himalayas, at Gaumukh on the

Indian side of the Tibetan border. It is 1560 miles (2510 km) long and flows through China, India, Nepal and Bangladesh. The Ganges River basin is one of the most fertile and densely populated in the world and covers an area of 400 000 sq miles (1,000,000 sq km). The river flows through 29 cities with population over 100,000, 23 cities with population between 50,000 and 100,000, and about 48 towns.

Pollution of the Ganges has become so serious that bathing and drinking its water has become very dangerous. The major polluting industry along the Ganges is the leather industry especially near Kanpur, from which chromium and other chemicals leak into the river. Another huge source of pollution is that of the nearly 1 billion litres of mostly untreated raw sewage that enters the river every day. Inadequate cremation procedures result in partially burnt or unburnt corpses floating in the river. The principal sources of pollution of the Ganga River can be characterized as follows: Domestic and industrial wastes, industrial sewage, solid garbage, agricultural run-off containing pesticides and fertilizers, defecation, mass bathing and ritualistic practices.

The Ganga Action Plan (GAP) was set up in 1985 by the Indian government with British and Dutch support to build a number of waste treatment facilities. Under the GAP, sewage is intercepted and the water is diverted for treatment and several electrical crematoria have been built. The broad aim of the GAP was to reduce pollution and to clean the river and to restore water quality at least to bathing quality. This was considered as a feasible objective and because a unique and distinguishing feature of the Ganges was its widespread use for ritualistic mass bathing. The other environmental benefits envisaged were improvements in fisheries, aquatic flora and fauna, aesthetic quality, health issues and levels of contamination. The major objectives were to improve the water quality, by controlling municipal and industrial wastes. The long-term objectives were to improve the environmental conditions along the river by suitably reducing all the polluting influences at source.

5. Mumbai : the dream city but polluted waters

Mumbai is a metropolitan with posh homes, slums; luxury cars, crowded trains; cramped factories and spectacular beaches. It is a city of dreams, attracting millions who come to earn fame, fortune and to make a living. In the last few decades, the city has witnessed major land reclamations, supported rural migration, and has struggled to expand basic services to cover the urban population. Still almost half of the city's 12 million residents are either in slums or homeless. The city generates 2,225 million litres of sewage per day that flows directly into the Arabian Sea.

Although, Mumbai is one of India's wealthiest cities in terms of human and financial capital, it is unplanned and the mismanaged urban growth threatens human health and well-being. Attempts by the Municipal Corporation of Mumbai (MCBM) to adequately dispose of sewage have been problematic due to shortages of land for secondary treatment, and by limited technical and financial resources. According to World Health Organization estimates, 80 percent of diseases in India are caused by water-borne pathogens. Cholera, salmonellosis, malaria, typhoid, and gastroenteritis are a few to name. The beaches, for which Mumbai was once famous, are choked with solid waste and raw sewage. Dissolved oxygen levels, an indicator of organic pollution in water, are seriously low along the most polluted part of the coast in Malad and Mahim, and moderately depleted along the entire coastline. Biochemical oxygen demand and nutrient levels are moderate in most of the region, and inorganic pollution such as metals is insignificant, but bacterial contamination is sometimes 100 times the permitted level.

To help deal with these massive environmental health challenges, the $300 million Bombay Sewage Disposal Project, funded by the MCBM and the World Bank was approved in 1995 to help the city invest in infrastructure that would collect and dispose 80 percent of the sewage currently flowing untreated into the sea.

6. Punjab : rich agricultural production but pesticide-laced

The Centre for Science and Environment (CSE), New Delhi conducted a study by randomly analyzing the blood samples of some people in the Punjab region. The results indicated very high levels of multiple pesticides (organochlorine, organophosphorous) in the individuals. This could have been due to direct exposure to the pesticides during application on the crops, or through ground water infiltration or food contamination. The Punjab Pollution Control Board reported a high incidence of the pesticide levels in vegetables and ground water. Pesticides are non-biodegradable toxins and remain persistent in the environment for long periods causing acute toxicity. Their bioaccumulation in the food chain causes severe toxicity. An alternate approach to the pesticide problem would be the use of biopesticides and implementation of biomonitoring practices.

7. The Minimata disease: A classic case study of inorganic pollution

Mercuric chloride was used as a catalyst in the production of plastics in the Chisso factory, a vinyl chloride factory, built on the Minamata Bay of Kyushu, Japan and the inorganic by product - inorganic mercury was dumped into the bay. Between 1953 and 1961, scientists discovered that the fishes in the bay were lethargic and some were dead. The studies confirmed that inorganic mercury had been converted to organic methyl mercury by methylation process carried out by the metabolism of the anaerobic microorganisms present in the bay waters. This organic mercury was toxic and bioaccumulated in the fishes and incorporated into the food chain. When humans consumed these fishes, it caused the Minamata disease. Paralysis, hearing and vision disorders, numbness, mental disorders, teratogenic effects, cellular distortions, cerebral palsy, etc characterized it. Similar case studies were reported in Japan when people consumed rice contaminated with cadmium, causing the Itai-itai disease, resulting in crippling of the legs.

Pollutants, their origin and effects: river/lake/ ground water pollution

The various water bodies (surface, ground waters) can be polluted from different sources. River (flowing water) and lakes (standing water) are surface water bodies, whereas ground water is underground water. There are two broad categories of water pollution. They are point sources, which occur when harmful substances are emitted directly into a water body, and nonpoint sources where pollutants indirectly enter water bodies through transport. The major sources include the following: industries / commercial (toxic compounds, inorganics, organics, heat, bleaching agents, etc), agricultural run off (pesticides, herbicides, insecticides, fertilizers, etc), domestic (detergents, sewage, oxygen demanding wastes, etc), oil spillages (tankers, deep sea drilling operations, etc) and natural geological sources (weathering, volcanic activities, etc). The following constituents characterize water bodies thus contaminated:

1. Inorganics

Common sources include industrial, commercial areas, mining, agricultural sources, weathering, domestic discharges, storm water, infiltration from cracked pipes and leaky manhole covers. Inorganic minerals such as sodium, potassium, copper, lead, nickel, zinc, calcium, iron, aluminium, magnesium, cadmium, are common in wastewater. Most inorganic substances are relatively stable and are not biodegradable. They contaminate soil and water. Some are toxic to animals and humans they accumulate in the environment resulting in bio-concentration and bioaccumulation. Inorganic ions in solution also contribute to conductivity (the ability to carry an electrical current) of the wastewater.

Some important inorganic pollutants are:

Mercury: It enters the water through industrial effluents (paper, plastic, battery industries, fungicides etc.). It had caused Minamata disease that resulted in deaths in Japan due to consumption of mercury contaminated fish.

Cadmium: It enters the water bodies through industrial discharge (electroplating, refining, welding, pesticides and fertilizer industries). It accumulates in the human liver and kidney leading to carcinogenesis.

Lead: This enters through industrial discharges (battery, lead pipes) and mining activities. It causes damage to liver, kidney, mental retardation, affects the central nervous system resulting in coma and death, abnormalities in fertility and pregnancy.

Fluorides: Naturally enters the water bodies due to geological factors, aluminium factories, and phosphate fertilizer plants. At higher concentration (more than 1.5 ppm), it causes fluorosis affecting the bones and teeth.

Arsenic: The major sources include its usage in pesticides and geological factor. Arsenic poisoning results in arsenicosis, which results in melanosis (mottling the skin), finally becoming leprosy like skin lesions and gangrenous ulcers.

2. Nutrients

Wastewater often contains large amounts of nutrients like nitrates and some amount of phosphates that promote plant growth. Nutrient pollution is a major problem in estuaries and deltas, where the runoff that was aggregated by watersheds is finally dumped at the mouths of major rivers.

Some important nutrients are:

Phosphorus

Phosphorus exists in wastewater in many forms and includes soluble orthophosphate ion (PO_4^{-3}), organically bound phosphate, and other phosphorus/oxygen forms. Most of the organic phosphates in wastewater are from excreta, domestic wastes containing phosphate detergents and food residues. Phosphorus is the limiting nutrient in freshwater surface waters and is the principal cause of eutrophication in surface water bodies like lakes. In acute cases, excess nitrogen and phosphorous can result in eutrophication leading to the nutrient enrichment of water bodies causing excessive growth of aquatic plants (algae, cyanobacteria, rooted aquatic vegetation, duckweed etc). They deplete the dissolved oxygen, release toxins and obnoxious odours and cause the death of organisms.

Nitrogen

Nitrogen comes from nitrogenous fertilizers used in agricultural fields and subsequently enters the ground water system. Nitrates in drinking water may contribute to miscarriages and is the cause of a serious illness in infants called methemoglobinemia or "blue baby syndrome."

3. Pesticides

Pesticides like DDT, cyanide compounds, methyl mercury enters into the wastewaters mainly through agricultural sources. DDT, a non-biodegradable organic is not soluble in water and it accumulates in the bodies of organisms. When higher organisms consume these organisms, the DDT becomes biomagnified as the concentration of DDT accumulates at each trophic level of the food chain. The process is often referred to as biomagnification.

4. Polycyclic aromatic hydrocarbons (PAHs)

These persist in the marine environment for several years and are extremely toxic to marine life at concentrations in the lower parts-per-billion range. Chronic and acute exposure to PAHs can affect development of marine organisms, increase susceptibility to diseases, and interfere with the normal reproductive cycles in many marine species.

5. Oil and grease

Oil and grease includes fats, oils, waxes, etc. Fatty organic material from animals, vegetables, and petroleum are not easily broken down by bacteria and can cause pollution in environments. They increase the BOD levels, and may float to the surface and harden, causing unpleasing conditions. They can trap plants and other materials causing foul odors and attract flies, mosquitoes and other disease vectors. Petroleum-based oils used in motors and industries are considered hazardous waste and they should be collected and disposed of separately from wastewaters.

6. Organic matter

Organic substances are oxygen demanding wastes, which are substances that oxidize in the water body and reduce the available amount of dissolved oxygen. Due to reduction in dissolved oxygen the aquatic life is threatened. They are usually biodegradable organic substances and originate from food and paper industrial effluents and municipal wastewaters. The commonly used measure of oxygen demand is (1) chemical oxygen demand – the amount of oxygen needed to chemically oxidize the wastes and (2) biological oxygen demand – the amount of oxygen used by microorganisms to biologically degrade the wastes.

Organic compounds normally are some combination of carbon, hydrogen, oxygen, nitrogen, and other elements. They are composed of the carbon-based chemicals that are the building blocks of most living things. Many organics are proteins, carbohydrates, or fats and are biodegradable, which means they can be consumed and broken down by organisms. Organic materials in wastewater originate from plants, animals, or synthetic organic compounds, and enter wastewater from human wastes, paper products, detergents, cosmetics, foods, agricultural, commercial, and industrial sources.

7. Pathogens

Common disease-causing viruses, parasites, and bacteria are present in wastewaters and are ubiquitous. These pathogens often originate from people and animals infected with diseases the sources include hospitals, schools, farms, and food processing plants. Gastroenteritis can result from a variety of pathogens in wastewater, and cases of illnesses caused by the parasitic protozoa *Giardia lambia* and other important wastewater-related diseases include hepatitis A, typhoid, polio, cholera, and dysentery. Outbreaks of these diseases can occur as a result of drinking water from wells polluted by wastewater, eating contaminated fish, or recreational activities in polluted waters.

8. Solids

Solid materials in wastewater can consist of organic (plant fibres, bacteria, algae) and/or inorganic materials (clay, silt, soil constituents) and organisms. They include total solids, suspended solids, dissolved solids, settleable solids. Solid analysis is an important pre-requisite for the control and treatment of wastewaters.

9. Temperature

Hot water, a byproduct of many manufacturing processes, can be a pollutant. Also nuclear plants release enormous heat and if this is released into a local stream or river the temperature rise can increase metabolic rates of organisms and drastically affect or even kill aquatic life. The temperature of wastewater is usually higher than that of the water supply because of the addition of warm water from domestic / industrial usage. When discharged in large quantities, it can raise the temperature of receiving streams locally and disrupt the natural balance of aquatic life.

10. pH and conductivity

pH is the measure of the activity of hydrogen ions and is given in logarithmic scale. The acidity or alkalinity of water affects the environment. Lower pH values than 7 indicates acidity, while pH greater than 7 indicates alkalinity (a pH of 7 is neutral).

The total amount of dissolved salts present in water can be easily estimated by measuring the specific conductivity of water.

11. Colour and turbidity

Colour and turbidity are aesthetically undesirable and impose extra loads on downstream water treatment plants and inhibit the self-purification. Microorganisms, clay, silt, metal oxides, dyes from various industrial, domestic, mining and refining processes can contribute to colour and turbidity in water bodies.

12. Odour

Industrial waters may contain compounds that produce objectionable odours. Again organic waters contain sulphides, sulphates etc. produced by microbial metabolism which may contribute to foul smell.

13. Gases

Certain gases in wastewater can cause odors and could be potentially dangerous. Methane gas, a byproduct of anaerobic biological treatment (found in septic tanks) is highly combustible. Hydrogen sulfide and ammonia gases can be toxic and can result in asphyxiation hazards.

Ammonia as a dissolved gas in wastewater is dangerous to aquatic life.

14. Volatile organic compounds

They are very common contaminants in ground waters. The most toxic VOC is vinyl chloride used in the production of polyvinyl chloride resins. Other VOCs include carbon tetrachloride used in fumigants, household-cleaning agents; tetrachloroethylene used as a solvent and in the manufacture of CFCs. They are very toxic and even a few milliliters can cause death. It is insoluble in water, a potent carcinogen, and causes injury to the central nervous system, kidneys and liver.

15. Oil spills

Oil spills like the Exxon Valdez spill, off the coast of Alaska (1989) and the more recent Prestige spill, off the coast of Spain have caused major water pollution problems for local aquatic, wildlife, fishermen, and coastal businesses. Shipping is one of the major sources of oil pollution in water. Again the discharge of oily wastes, oil-contaminated ballast water, wash water, drilling and extraction operations for oil and gas are all significant sources of marine pollution.

Water quality standards

Drinking water problem is not only in terms of availability of quantity but also quality of the water available. The reasons for chemical and bacteriological contamination are: poor hygienic conditions around the water sources, improper disposal of sewage and industrial waste water, callous disposal of solid waste, indiscriminate use of chemical fertilizers having high quantity of nitrates used in the agricultural sector, pollution from industrial effluents (untreated), over-exploitation leading to quality degradation, pollution of the source due to ignorance of the people, over-population and lack of public awareness.

Quality Criteria of Drinking Water : As Prescribed by Indian Standards Institution and Indian Council of Medical Research

1. Standards of bacteriological quality

i) Water in distribution system

- Throughout year, 95% of samples tested should not contain any coliform organisms (Test sample should be of 100 ml) and no sample should contain more than 10 coliform organisms per 100 ml of test sample.
- Coliform organisms should not be detectable in 100 ml. of any two consecutive samples.

ii) Unpiped water supplies

Where it is impracticable to supply water to consumers through a piped distribution network and where untreated sources, such as wells, boreholes and springs which may not be naturally pure, have to be used, the requirements for piped supplies may not be attainable. In such circumstances disinfection is most desirable and considerable reliance has to be placed on sanitary inspection and not exclusively on the results of bacteriological examination.

2. Standards of physical & chemical quality

Table 4.3 : Physical and chemical quality of drinking water

S. No. (1)	Substance or Characteristic (2)	Prescribed By ISI requirement (desirable limit) max. permissible level (3)	Prescribed by ICMR highest desirable level (4)	Maximum permissible level (5)
1.	Colour, (Hazen units)	10 units	5 units	25 units
2.	Odour	Unobjection-able	Unobjection-able	Unobjection-able
3.	Agreeable	Unobjection-able	Unobjection-able	Unobjection-able
4.	Turbidity	10 NTU	5JTU	25 JTU
5.	Dissolved solids, mg/l	500	500	1500*
6.	pH value	6.5 to 8.5	7.0 to 8.5	6.5 to 9.2
7.	Total hardness (as CaCO3), mg/l	300	300	600
8.	Calcium (as Ca), mg/l	75	75	200
9.	Magnesium (as Mg), mg/l	30	Not more than 50 mg/l, if there are 200 mg/l sulphate; if there is fewer sulphates, magnesium up to 100 mg/l, may	

Contd...

Contd...

(1)	(2)	(3)	(4)	(5)
			be allowed at the rate of 1mg/l for every 4 mg/l decrease in sulphates	
10.	Copper (as Cu), mg/l	0.05	0.05	1.5
11.	Iron (as Fe), mg/1	0.3	0.1	1.0
12.	Manganese (as Mn), mg/1	0.1	0.1	0.5
13.	Chlorides (as Cl), mg/l	250	200	1000
14.	Sulphate (as SO4), mg/l	150	200	400
15.	Nitrate (as NO3), mg/l	45	20	**
16.	Fluoride (as F), mg/l	0.6 to 1.2	1.0***	1.5
17.	Phenolic Compounds (as C6H5OH), mg/l	0.001	0.001	0.002
18.	Mercury (as Hg), mg/l.	0.001	—	0.001
19.	Cadmium (as Cd), mg/l	0.01	—	0.01
20.	Selenium (as Se), mg/l	0.01	—	0.01
21.	Arsenic (as As), mg/l	0.05	—	0.05
22.	Cyanide (as CN), mg/l.	0.05	—	0.05
23.	Lead (as Pb), mg/1.	0.10	—	0.10

Contd...

Contd...

(1)	(2)	(3)	(4)	(5)
24.	Zinc (as Zn), mg/l	5	—	—
25.	Anionic detergents (as MBAS), mg/l	0.2	—	—
26.	Chromium (as Cr6+), mg/1	0.05	—	—
27.	Polynuclear aromatic hydrocarbons (as PAH), mg/1	-	—	—
28.	Mineral oil, mg/1	0.01	—	—
29.	Residual free Chlorine, mg/1.Min.	0.2	—	—
30.	Pesticides	Absent	—	—
31.	Radio-active materials:		—	
	a) Alpha emitters uc**** per ml, Max.	10-8	—	3 pci/l*****
	b) Beta emitters uc per ml, Max	10-7	—	30 pci/1

*Dissolved solids relaxable upto 3000 mg/1 in case where alternate sources are not available within reach.

** More information is required to prescribe a value but in no circumstances should the level exceed 100 mg NO_3

*** The presence of fluoride in drinking water in excess of 1.0 mg/1 gives rise to dental fluorosis (mottling of varying degrees of severity in children. When present in high concentrations, fluorides may eventually cause endemic cumulative fluorosis with resultant skeletal damage in children and adults

**** uc – micro Curie

***** pCi – 'Picocurie'- unit of radio-active disintegration, also known as micro microcuric equivalent to disintegrations per second

Control of Water Pollution

1. Minimize the use of pesticides, herbicides, insecticides on agricultural lands.
2. Construction of bunds to prevent run off from agricultural fields
3. Proper treatment of waste waters before discharge into natural water bodies
4. To use bio-pesticides, bio-fertilizers, bio-monitoring, adopting integrated pest management practices
5. Planting more trees to limit soil erosion and run off

Waste Water Treatment

The waste water standards are designed to control pollution by setting limits on the quality of the discharged wastewater. It establishes basic effluent limits for all facilities to protect surface water quality standards. These wastewater permits regulate the quality of wastewater that is discharged into a stream, river, lake, or bay. The permit levels are a reflection of the state water quality standards that need to be met for specific water bodies. The Water prevention and control of pollution act, 1974 (India) amended in 1991 was aimed at restoring, maintaining the quality of water bodies and to prevent pollution of waters.

Wastewater or contaminated water is normally called as influent when it enters the wastewater treatment plant. Here, the water is treated before it is returned back to the environment (lakes, or streams or ponds). Most treatment plants have primary, secondary and tertiary treatments (Fig 4.2). Sewage treatment is a multi-stage process and the purpose is to clean the water i.e. to reduce and remove organic matter, solids, nutrients, and disease causing organisms and other pollutants from wastewater. Water thus contaminated with undesirable solids, microorganisms, inorganics, organics etc is treated in wastewater treatment plants. There are a number of processes and techniques used in the treatment procedures.

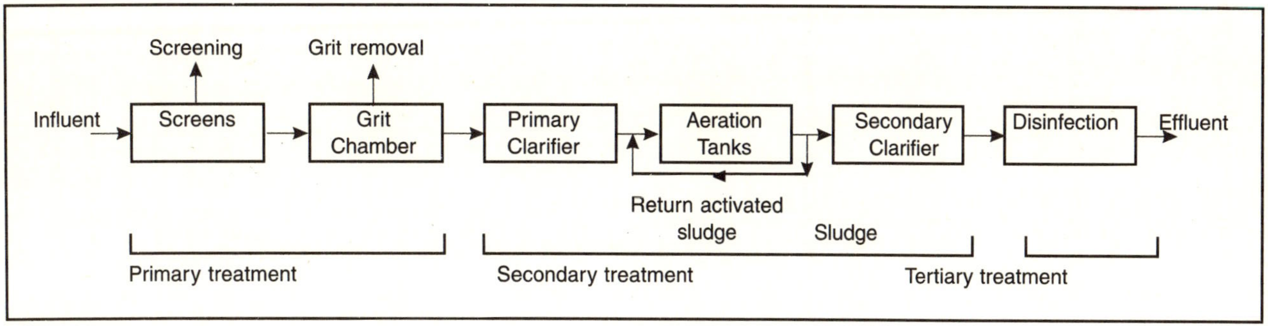

Fig. : 4.2 — Schematic flow chart for wastewater treatment

Treatment objectives

The selection of technology ultimately depends upon wastewater characteristics and on the treatment objectives. The primary objective of wastewater treatment and disposal is to prevent pollution to waters whether they are rivers and canals, ponds and lakes, or tidal estuaries and coastal waters. The type of treatment and effluent quality control is typically aimed at public health protection (recreation, irrigation, water supply), preservation of the oxygen content in the water, prevention of eutrophication, preventing toxic compounds from entering the water and food chains, and promotion of water reuse.

Table 4.4 : Classification of some important waste water treatment technologies

Primary	Secondary	Tertiary	Advanced
1. Screens 2. Grit chambers	1. Aerobic – ➢ Trickling filters ➢ Rotating biological contactors (RBC) ➢ Activated sludge ➢ oxidation ditch ➢ Lagoon systems 2. Anaerobic-USAB	1. Disinfection 2. Chemical precipitation	1. Chemical Oxidation 2. Catalytic wet air oxidation 3. Ultra violet enhanced oxidation

1. Primary treatment

It is the first stage of wastewater treatment process that removes solids that settle down or float. It generally removes upto 40% of the suspended solids and 30-40% of the BOD in the wastewater. Wastewater is held in a sedimentation tank for several hours allowing the particles to settle down to the bottom and the oil, greases to float on the surface. The sedimentation tank is a tank where solids settle down by gravity. The settle-able solids settle down and are pumped out, while oils float on the surface and are skimmed off. Clarifiers and septic tanks are usually used to provide primary treatment. This involves the following steps.

- **Screening-** removes large objects, such as stones or sticks that could block tank inlets.
- **Grit chamber-** a chamber or tank where wastewater slows down and heavy, large solids called grit settle out and are removed

2. Secondary treatment

The water obtained from the primary treatment is treated using biological treatment processes (activated sludge, trickling filters, etc.) followed by settling tanks. This step removes approximately 85% of the BOD and total suspended solids producing a relatively highly treated effluent. Secondary treatment is essentially a biological treatment process to remove dissolved organic matter from wastewaters. Sewage microorganisms are cultivated and added to the wastewater. The microorganisms absorb organic matter from sewage as their food intake. Approaches to secondary treatment include aerobic (fixed film, suspended film and lagoon systems) and anaerobic systems.

i. Aerobic systems

A. Fixed film systems

Fixed film systems grow microorganisms on substrates such as rocks, sand or plastic. The wastewater is spread over the substrate, allowing the wastewater to flow past the film of microorganisms fixed to the substrate. As organic matter and nutrients are absorbed from the wastewater, the film of microorganisms grows and thickens. Trickling filters, rotating biological contactors, and sand filters are examples of fixed film systems.

➤ Trickling filters

It is a biological treatment process using coarse media like rock or plastics that are contained in a tank, which serves as a surface on which microbiological growth occurs (Fig 4.3). Wastewater trickles over the media and microorganisms helps in removing the pollutants (there by reducing BOD and TSS). Settling tanks to remove microorganisms that wash off or pass through the trickling filter media follows trickling filters. These are beds of coarse media 3-10 ft. deep. Air is mixed with wastewater (aeration), and then allowed to trickle through the media. Microorganisms, attached and growing on the media break down organic material in the wastewater. Trickling filters drain at the bottom; the wastewater is collected and then undergoes sedimentation.

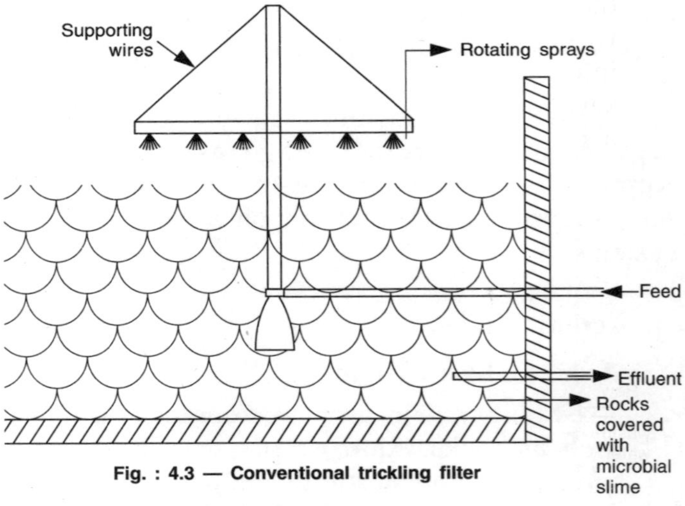

Fig. : 4.3 — Conventional trickling filter

➤ Rotating biological contactors

This process involves wastewater to come in contact with a biological media to facilitate the removal of contaminants. A rotary biological contactor consists of a series of discs mounted on a shaft, which is driven so that the discs rotate at right angles to the flow of settled sewage (Fig 4.4). The discs also called bio discs are usually made of plastic (polythene, PVC, expanded polystyrene) and are contained in a trough so that about 40 percent of their area is immersed in wastewater. The discs are arranged in groups or packs with baffles between each group to minimize surging or short-circuiting. The units are usually installed in a concrete tank so that the surface of the wastewater passing through the tank almost reaches the shaft. The shaft continually rotates at 1 to 2 rpm, and a layer of 2 to 4 mm thick biological growth is soon established on the wet surface of each disk. The biological growth (algae, cyanophyceae, chlorella etc) that becomes attached to the disks assimilates the organic materials in the wastewater. Aeration is done by the rotating action, which exposes the disks to air after contacting them with the wastewater. Excess biomass is sheared off in the tank, where the rotating action of the disks keeps the solids in suspension. Eventually, the flow of the wastewater carries these solids out of the system and into a clarifier, where they are separated. By arranging several sets of disks in series, it is possible to achieve a high degree of organic removal and nitrification.

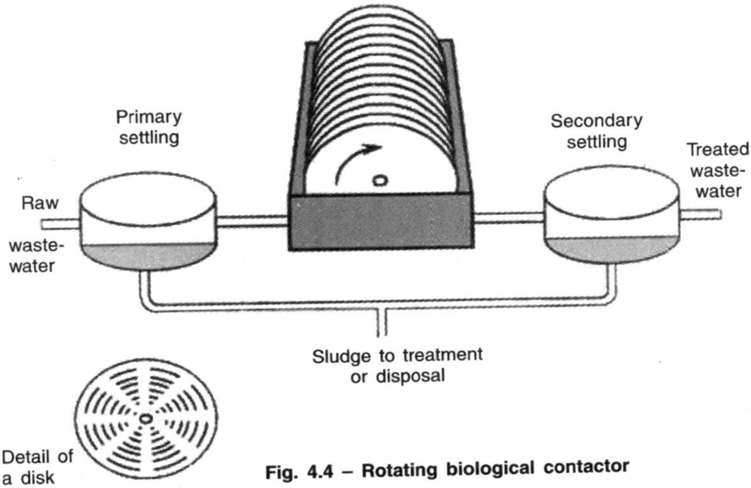

Fig. 4.4 – Rotating biological contactor

B. Suspended film systems

Suspended film systems agitate and suspend microorganisms in wastewater. As the microorganisms absorb organic matter and nutrients from the wastewater they grow in size and number. After the microorganisms have been suspended in the wastewater for several hours, they are settled out as sludge. A part of the sludge is pumped back into the incoming wastewater to provide "seed" microorganisms. The remainder waste is sent to a sludge treatment plant. Activated sludge, extended aeration, oxidation ditch, and sequential batch reactor systems are some examples of suspended film systems.

➢ **Activated sludge**

The method primarily uses microorganisms in the treatment process to break down organic material by aeration and agitation (Fig 4.5). The biological treatment is carried out by the interaction of different types of aerobic microorganisms, which grow and multiply in order to degrade the organic matter. This results in the removal of dissolved organic matter and nutrients from the wastewater. The

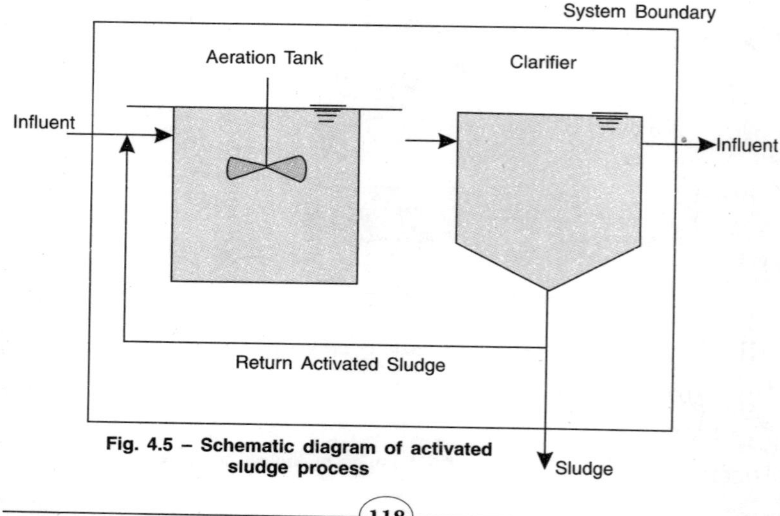

Fig. 4.5 – Schematic diagram of activated sludge process

resulting sludge from the above process is called activated sludge. The activated sludge exists normally in the form of flakes, which contains living and dead biomass. Bacteria containing "activated sludge" are continually recirculated back to the aeration basin to increase the rate of organic decomposition. The activated sludge flows from the biological aeration basin into the final clarifier where the flakes settle down to the bottom and can be separated from the cleaned wastewater. The main part of the separated sludge, which is transported back to the aeration basin, is called return activated sludge.

- **Oxidation ditches**

It consists of a long channel elliptically shaped equipped with an aeration equipment called a rotor for generating a water flow and stirring water in the channel to supply oxygen, a settling tank for collection of sludge which is recirculated back to the ditch. The oxidation ditch is used for long-term aeration. Although it requires a relatively large area, it has a simple structure and can be easily operated and it can remove nitrogen easily.

- **Aerated lagoons**

Lagoon systems are shallow basins that hold wastewater for several months, allowing natural degradation of sewage. These systems use natural aeration and microorganisms in the wastewater to renovate sewage. These are slow, cheap, and relatively inefficient and rely on the interaction of sunlight, algae, microorganisms, and oxygen.

ii. Anaerobic systems

Up flow Anaerobic Sludge Blanket Reactor. (UASB)

It is also known as the anaerobic contact process as it occurs in the absence of oxygen and is highly efficient in the removal of organic content of the wastewater. Up flow Anaerobic Sludge Blanket (UASB) process is the most cost-effective in industrial wastewater treatment. The anaerobic process achieves 75-85 per cent of the effluent standards, minimal N and P removal and a better-stabilized sludge.

UASB reactors

The wastewater flows through the inlet and then into the Flow Equalization Tank (FET), which is a deep concrete storage. There are three agitators placed inside the FET that gently mix the influent to homogenize it. This prevents settling of suspended solids. Due to the storing of the influent in the tank, a reduction in pH occurs through acidogenic fermentation. Acidogenesis is a process where dissolved compounds are converted into simple compounds, (volatile fatty-acids, alcohols, lactic acid, CO_2, H_2, NH_3, H_2S). Studies suggest that a detention time of 6 hrs is optimum to achieve the desired pre-acidification.

The wastewater flow from the FET is pumped to the UASB through the USAB feed pumps. The reactor consists of an empty volume covered with a plate settler zone to collect and recycle suspended matter escaping from the sludge blanket below. The water flows upwards through a blanket of suspended granules containing the active biomass. When the influent flows through the loops and enters the reactor chamber, hydrolysis reaction occurs. Hydrolysis is a process where enzymes excreted by fermentative bacteria convert complex, heavy, undissolved matter (proteins, carbohydrates, fats) into less complex, lighter, materials (amino acids, sugars, alcohols). As the flow rises through the granulated sludge, the hydrolysis reaction is converted into an Acetogenesis reaction. Acetogenesis is a process where digestion products are converted into acetate, H_2, CO_2 and new cell-matter. As the rise continues, the acetogenesis is finished and it is converted to a methanogensis phase. Methanogenesis is a process where acetate, hydrogen plus carbonate, formate or methanol, are converted into CH_4, CO_2 and new cell-matter. COD removal of up to 80% has been achieved from the influent. The methane and CO_2 bubbles are caught below the plate settlers and taken out of the reactor separately.

3. Final treatment or tertiary treatment

Final treatment involves the removal of disease-causing organisms from wastewater. After primary and secondary treatment, municipal wastewater is usually disinfected with chlorine or other disinfecting compounds, ozone or ultraviolet light. Sludge from primary and secondary treatment settling tanks are given further treatment and undergo several options for disposal.

4. Advanced treatment

Advanced treatment is necessary in some treatment systems to remove nutrients from wastewater. It is an advanced level of treatment that uses filtration, nutrient removal (removal of nitrogen and phosphorus), carbon adsorption to remove chemicals and the removal of toxic chemicals or metals. These processes can be physical, biological, or chemical. Some examples of nutrient removal systems include coagulant addition for phosphorus removal and air stripping for ammonia removal. Advanced chemical oxidation processes make use of chemical oxidants to reduce COD/BOD levels, and to remove both organic and oxidisable inorganic components. The processes can completely oxidize organic materials to carbon dioxide and water. The types of advanced oxidation processes are:

- Chemical oxidation processes using hydrogen peroxide, ozone, combined ozone & peroxide, hypochlorite, Fenton's reagent etc.
- Ultra-violet enhanced oxidation such as UV/ozone, UV/hydrogen peroxide, UV/air
- Wet air oxidation and catalytic wet air oxidation (where air is used as the oxidant)

Advanced oxidation processes has advantages over biological or physical processes, including as it can be operated unattended, there is absence of secondary wastes and it can handle fluctuating flow rates and compositions. It is appropriate for effluents containing refractory, toxic or non-biodegradable materials. However, advanced oxidation processes often have higher capital and operating costs compared with biological treatment.

Some Interesting Facts

- In India the main water consuming activities are bathing and washing clothes. The water wastage is highest in washing (82%), bathing (52%) and washing utensils (49%).
- The major countries affected by arsenic poisoning in ground waters are Chile, Argentina, India, Taiwan, Mexico, China and Bangladesh.
- 17 million residents in Sri Lanka are prone to dental fluorosis due to excess fluoride in the drinking waters. The waters come from naturally fluoride rich ground water affected by the chemistry and mineralogy of the soils. Dental fluorosis (mottling of teeth) occurs when the fluoride levels exceed 1.5 milligrams per litre.
- Extensive pumping of groundwater can cause land subsidence. In China, more than 20 cities have been seriously affected by land subsidence due to uncontrolled groundwater withdrawal.
- The use of enzymes (Xylanase) in the paper manufacturing industries reduces the use of bleaching agents like chlorine by around 15%.
- Aquaculture accounts for approximately 35% of the world's fishmeal consumption. It produces large amounts of nutrients and organic wastes and can pollute water bodies if not managed properly.

UNIT 5

LAND POLLUTION

Land is an indispensable resource and we depend on it for our food, fuel, clothes and energy. Hence, it should be properly managed. Land degradation can be natural and anthropogenic / man-made. Natural sources are those that occur spontaneously in nature, due to the action of wind, landslides, rainfall etc. Man-made activities that cause degradation include: over extraction of ground water leading to subduction or subsidence of land, overgrazing by animals, deforestation, improper waste disposal mechanisms, mining of resources, burning, etc. Land erosion varies with soil types and depends on topography, altitude, wind intensity, rainfall etc. Degradation of soil is much faster than formation. For example, the annual erosion rates for agricultural lands are approximately 20 – 100 times the average of 500 years required to renew 2.5 cm of new soil in a tropical region. Again, soil erosion leads to vital nutrients being lost from the soil, alkalinization and it leads to desertification, decline in biomass, decline in agricultural production and decrease in water tables. The land is a part of the lithosphere, which is the upper part of the earth's crust.

Czech Republic: A Beautiful Country Defaced by Severe Uranium and Coal Mining

The coal deposits in North Bohemia, Czech Republic is mined extensively for the raw materials required for energy production. The whole country is naturally beautiful with mountains and ancient churches and architecture. Coal mining has destroyed many villages, destabilized hillsides and made the land barren and difficult for recultivation. Emissions from power plants have acidified the soils and surface waters. 60 kg of SO_2 is released into the air for every ton of coal burned. The spruce forests in the region are totally declining. Another important mine in the Czech is the uranium mines. Effective leaching of the uranium ores requires 10 – 15 years for the necessary solution – rock interaction to release uranium. Due to this the ground waters have been replaced by a highly concentrated sulphuric acid solution that percolates through adjacent clay sediments to pollute aquifers and local streams. This would seriously pollute the Ploucnice River by 2010. An estimate of about $1.6 billion alone is needed to restore the regions affected by uranium mining. The resulting coal and uranium mining has caused severe pollution and degradation of the soils; air and water as harmful chemicals are released from coal and uranium mines.

The above case study shows that the environmental concerns were totally ignored and the gorgeous country now looks like a lunar landscape. Designing effective methods of land restoration requires careful studies on the geology, geochemistry and hydrology of the region.

Lithosphere

The lithosphere (in Greek means "rocky" sphere) is the solid outermost shell of our planet and it includes the crust and the uppermost layer of the mantle. The Earth is an oblate spheroid composed of different layers as determined by deep drilling seismic studies and geophysical studies. These layers are: crust, mantle, inner and outer core. The distinguishing characteristic of the lithosphere is not composition, but its flow properties. The movement of lithospheric plates is known as plate tectonics.

Soil

Soil is the uppermost layer of the lithosphere. It means *solum* in Latin meaning earthy material. Soil originates with the weathering of bedrock and the reorganization of this material by water, plants, animals and detritivores. Humus is the decomposed component of soil. Topsoil takes thousands and even millions of years to form; soil erosion is a serious loss of resources.

Weathering and soil formation

Weathering is the breakdown and alteration of rocks and minerals at or near the Earth's surface into products that are more in equilibrium with the conditions found in the environment. It is the first step for a number of geomorphic and biogeochemical processes. The products of weathering are a major source of sediments including sand, silt, and clay that contribute to the formation of soil. The release of ion salts from rocks and minerals make the oceans saline. Leaching and runoff transport these ions from land to the ocean basins where they accumulate in seawater. Weathering is a process that is fundamental to many other aspects of the hydrosphere, lithosphere, and biosphere. There are three broad categories of mechanisms for weathering: chemical, physical and biological.

➢ Chemical weathering

Chemical weathering involves the alteration of the chemical and mineralogical composition of the weathered material. A number of different processes can result in chemical weathering. The most common chemical weathering processes are hydrolysis, oxidation, reduction, hydration, carbonation, and solution.

- **Physical weathering**

  Physical weathering is the breakdown of mineral or rock material by entirely mechanical methods brought about by a variety of causes. Some of the forces originate within the rock or mineral, while others are applied externally. Both of these stresses lead to strain and the rupture of the rock. The processes that may cause mechanical rupture are abrasion, crystallization, thermal insolation, wetting and drying, and pressure release.

- **Biological weathering**

  Biological weathering involves the disintegration of rock and mineral due to the chemical, physical agents and chelation of plants and organisms.

Soil profile

A soil profile is the term used for describing the vertical section of the crust upto the parent rock (Fig 5.1). Each horizon or layer has a different colour, thickness, texture, structure, porosity and composition. There are four basic horizons. They are:

- **O horizon**: It is the organic zone, consisting of freshly fallen litter, leaves, organic debris, which is partly or not decomposed. It is rich in life.

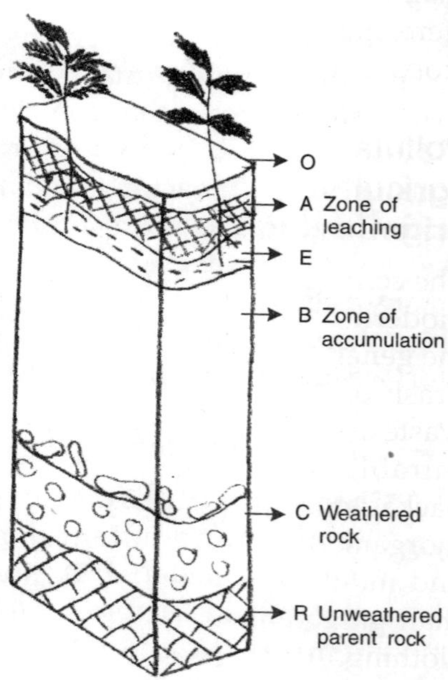

Fig. 5.1

- **A horizon** is the mineral rich layer and uppermost topsoil; it contains litter and humus but soluble chemicals have been leached out. It is lighter in colour.
- **B horizon** has little or no organic matter but contains inorganic nutrients leached from the A horizon above. It is rich in clay, humus, iron and aluminium. The A and B layers and often termed as the true soil.
- **C horizon** is the weathered or shattered rock and consists of weathered mineral constituents.
- **R horizon** is the unweathered parent rock.

There are different types of soils that include the sandy soil, clayey soils, loamy soils, clay loam soils and the silt loam soils. Grassland soil has a deep A horizon built up from decaying grasses over many years with little leaching into the B-horizon. Forest soils have a lot of inorganic nutrients in the A and B-horizons to allow root growth. In the tropical rain forests, the A horizon is very shallow and the B-horizon is deeper due to intense leaching. Here since the topsoil lacks nutrients, the soil can support crops only for a few years.

Pollutants (municipal, industrial, commercial, agricultural, hazardous solid wastes), their origin and effects

The combined effects of population explosion and changing modern living standards have had a cumulative effect in the generation of a large amount of various types of wastes. Trash or garbage is commonly known as municipal solid waste (MSW). MSW is normally defined as wastes including durable goods, non-durable goods, containers and packaging, food wastes, yard wastes, and miscellaneous inorganic wastes from residential, commercial, institutional, and industrial sources. Common examples include (other than gaseous and liquid wastes): appliances, newspapers, clothing, food scrapes, boxes, disposable tableware, paper, wood pieces, rubber tyres, and cafeteria wastes, product packaging, grass clippings, furniture, clothing, bottles, appliances, paint, and batteries etc. The solid wastes can also be hazardous, in which they are termed as hazardous waste.

India generated an estimated 6 million tonnes of solid waste in 1947 and in 50 years i.e. 1997 it increased to 48 million tonnes. In India, 25% of the municipal solid waste is not collected and 70% of the Indian cities lack facilities to transport it and there are few or no sanitary landfills to dispose of the waste. The existing landfills are neither well equipped nor lined properly to protect against soil and groundwater contamination.

For comparison of data on waste, in the U.S. in the year 2003, the residents, businesses, and institutions produced more than 236 million tons of MSW, which amounts to approximately 4.5 pounds of waste per person per day.

A Classic Case Study of Waste Mismanagement: The Love Canal tragedy

In 1892, a small area near Niagara Falls, New York, USA was dug up and designed as a route for transportation and for the production of hydroelectric power connecting the lower and upper reaches of the Niagara River, called the Love canal (named after William T. Love). The process was left incomplete and eventually became a site for waste disposal. A Plastic manufacturing chemical company disposed off its wastes in sealed steel drums during the years 1940 to 50. More than 20,000 tons of 80 different chemicals were dumped in the area. In 1953, the company sold the land to the city of Niagara Falls. Subsequently, many residences and an elementary school were built on the land. The tragic story of the love canal incident started after twenty years due to heavy snow and rains. The sealed plastic wastes had corroded the drums and started leaking into the soil, buildings, water etc. Children in the school were shocked to see their rubber shoe soles disintegrating and they also suffered burns. This caused the officials to look into the matter and it was found that the wastes contained lot of carcinogens and a wide array of toxic organic compounds like benzene, chloroform, dichloroethylene etc. The clean up of the Love canal totally amounted to about 275 million US dollars. The schools and homes were destroyed, the families relocated, the soil was completely covered with compacted clay, the wastes were diverted to a treatment plant, the water pipes were blocked with barriers, etc. This was a classic example of solid waste mismanagement.

Toxic materials pose significant health hazards and dwindle our precious bio-reserves. Proper and timely management of waste is very important in the good interest of our environment, and us. The above classic case study proves that management of wastes is a very serious business and if not handled with care, the cleaning up of improperly disposed wastes can cause loss/damage to human life and be extremely costly. It therefore makes sense to plan the disposal of various kinds of wastes in a scientific manner.

Types and sources of solid waste

Solid waste can be classified into different types depending on their source.

1) Municipal solid waste

Municipal solid waste is generated mainly from residential and commercial complexes. It consists of household waste, construction and demolition debris, sanitation residue, and waste from streets. The different types of domestic wastes generated and the time taken for them to degenerate is illustrated in the table given on next page.

Table 5.1 : Domestic wastes and their degeneration time	
Common domestic wastes	**Approximate time taken for degeneration**
Organic kitchen waste – vegetables, fruits	1–2 weeks
Paper, cardboard paper	15 days–1 month
Cotton clothes	2–5 months
Woollen clothes	about a year
Metal cans, tin, aluminium	100–500 years
Plastics	1 million years

2) Hazardous waste

Hazardous wastes are highly toxic to humans, animals, and plants. They are corrosive, highly inflammable, and explosive and react when exposed to gases. Hazardous wastes include old batteries, acids, explosive materials used in fire-crackers, shoe polish, paint tins, old medicines, formaldehyde, phenols, mercury etc. India generates 7 million tons of hazardous wastes every year, of which most of it is concentrated in Andhra Pradesh, Bihar, Uttar Pradesh, and Tamil Nadu.

3) Hospital waste

It is generated during surgical / diagnostic treatments, immunization of human beings and in the production or testing of biologicals. It may include wastes like sharps, soiled waste, disposables, anatomical waste, disposable syringes, swabs, bandages, body fluids, human excreta, cultures, discarded medicines, chemical wastes, etc. This waste is highly infectious and can be a serious threat to human health if not managed in a scientific and discriminate manner. It is roughly estimated that at least 1 kg of the waste generated in a hospital is infected out of approximately 4 kg generated. Most of the hospitals in India do not have proper disposal facilities. Health care establishments in

India are slowly streamlining the process of waste segregation, collection, treatment, and disposal after the notification of the Bio-medical Waste (Handling and Management) Rules, 1998 by the Ministry of Environment and Forests, Govt. of India.

Table 5.2 : Different sources of waste

Sources	Types of waste generated
Domestic / Residential	Food wastes, broken glassware, scraps, newspapers, pens, pencils, plastics, rubber-bands, cloths, diapers, paper, cardboard, textiles, leather, yard wastes, wood, metals, ashes, batteries, oil, tyres, etc.
Agriculture and horticulture	Fertilizer packaging materials, grass cuttings, stems, twigs, branches, rotten fruits, decayed flower and leaf debris, etc.
Construction sites	Wood, tiles, steel, cement blocks, stones, concrete, bricks, nails, etc.
Commercial	Paper, cardboard, plastics, wood, food wastes, glass, metals, special wastes, hazardous wastes.
Industrial	Industrial process wastes, scrap materials, tailings, acids, paper pulp wastes, rubber, electrical parts, packaging materials, boxes, batteries, inorganics, organics, liquid wastes, metals, dyes, asbestos, paints, chemicals, bulbs, spray cans, etc.
Hospitals / Clinics	Disposable syringes, cotton, gauze, infectious materials, medicine strips, drip bottles, old medicines, files, papers, liquid wastes, cloth soiled with blood and other body fluids etc.
Nuclear / Thermal	Radioactive wastes, flyash etc.
power plants Treatment plants	Effluents, residual sludge

Biodegradable waste can be degraded by microbial activity whereas non-biodegradable wastes cannot be degraded by microbial activity, as it is evident from the terms used.

Properties of Waste

It describes the individual constituents that make up a solid waste stream and their relative distribution, based on percentage by weight. Waste composition is important for evaluating equipment needs, systems, the design and potential of recycling systems, and for planning purposes.

1. Physical properties

They include density, moisture content, particle size and distribution, field capacity, and porosity.

➢ **Density**

It is the weight per unit volume and is expressed as kg/m^3. Density is an important parameter to be determined because it is needed to assess the total mass and volume of waste, to be properly managed. Density varies with variety of waste constituents, the degree of compaction, the state of decomposition and the total depth of waste.

➢ **Moisture content**

Moisture content is important in the decomposition processes, the flushing of inorganic components, and the use of MSW in incinerators. Pre-treatment of waste to ensure uniform moisture content can be carried out prior to landfill disposal.

➢ **Particle size and distribution**

The size and distribution of the components of wastes are important for the recovery of materials, especially when mechanical means are used, such as screens and magnetic separators.

- **Field capacity**

  The field capacity of MSW is the total amount of moisture, which can be retained, in a waste sample subject to gravitational pull. Water in excess of field capacity will form leachate, which can be a major problem in landfills.

- **Permeability of compacted wastes**

  The hydraulic conductivity of compacted wastes regulates the movement of liquids and gases in a landfill. Permeability depends on pore size distribution, surface area and porosity.

2. Chemical properties

The chemical composition of waste helps to evaluate alternative processing and recovery options. This is very important where wastes are burnt for energy recovery. The most important chemical properties include: fusing point of ash, elemental analysis, and energy content.

- **Fusing point of ash**

  This is the temperature when ash resulting from the burning of waste forms a solid by fusion. Typical fusing temperatures range between 1100 - 1200°C.

- **Elemental analysis**

  It involves the determination of carbon, hydrogen, oxygen, nitrogen, sulphur, and ash. It is important in determining nutrient availability, to characterize the composition of the organic matter in wastes and C/N ratios for biological decomposition.

- **Energy content**

  The energy content of the components of waste can be determined using a boiler system, laboratory bomb calorimeter, or calculation using elemental composition.

- **Essential nutrients**

  If the organic content of MSW is to be used for compost, methane or ethanol production, then it is important to know the essential nutrient content. The major nutrients of importance are nitrogen, phosphorus and potassium.

3. Biological properties

They include the organic fraction of MSW and can be classified as follows:

- **Water-soluble constituents** - sugars, starches, amino acids and various organic acids
- **Hemi cellulose** - a product of 5 and 6-carbon sugars
- **Cellulose** - a product of 6-carbon sugar glucose
- **Fats, oils and waxes** - esters of alcohols and long-chain fatty acids
- **Lignin** - present in some paper products
- **Lignocelluloses** - combination of lignin and cellulose
- **Proteins** - amino acid chains

Solid waste management

In the last decade, the Indian consumer market has grown rapidly leading to products packed in cans, plastic bottles, tins, aluminium foils, plastic bags / boxes and other innumerable non-biodegradable items that persist in the environment. Some villages and towns have banned the use of plastics and they have achieved tremendous success. An exquisite example is the entire district of Ladakh. The local authorities imposed a ban on plastics in 1998 and not a single piece of plastic is seen there today. The first state in India to ban polythene and plastic bags/ covers was Himachal Pradesh. Other cities and states should follow the example of Ladakh, Himachal Pradesh and ban the use of non-biodegradable items

Source reduction, recycling, reusing and composting, prevent or divert materials from the waste stream are the normal proper waste management practices. Source reduction involves altering the design, manufacture, or use of products and materials to reduce the amount and toxicity of materials that get thrown away. Recycling diverts items, such as paper, glass, plastic, and metals, from the waste stream. These materials are sorted, collected, and processed and then manufactured, sold, and bought as new products.

Man has not even spared the oceans for dumping the wastes and this is known as Ocean Dumping. The oceans cover more than 70% of our earth. They regulate the biogeochemical cycles and the climate. The marine environment has been severely polluted following the dumping of industrial wastes. The wastes include acids, paper mill wastes, pesticides, refinery waste, residual sludge, radioactive waste, garbage, sewage, tiles, stones, clay, rocks, etc. Marine pollution results in the death and retardation of marine birds, organisms, reduction in dissolved oxygen vital for organisms, eutrophication problems. The marine ecosystems like the coral reefs, estuaries, salt marshes, and mangrove swamps are seriously threatened. Waste management strategies and environmentally friendly alternatives could curb this problem. The integrated waste management (IWM) is defined as a set of management alternatives that include reuse, source reduction, recycling, composting, landfill and incineration.

Collection and disposal of solid wastes

It is important to know how much waste is generated for proper management strategies. This involves assessing and analyzing the amount of wastes generated at source and their appropriate separation for reuse, recycling and finally proper disposal mechanisms. The collection of garbage is the most expensive part of municipal waste management systems. Despite all the changes that have taken place, the collection and transfer of wastes accounts for about two-thirds to three-quarters of the costs of waste management systems. Waste collection means choosing the most appropriate trucks, designing collection routes, container types for waste collection and then administering the collection. The steps involved include the following:

- Analysis of waste generated
- Waste separation at source
- Waste collection
- Transformation, recycling, re-use of wastes
- Disposal mechanisms

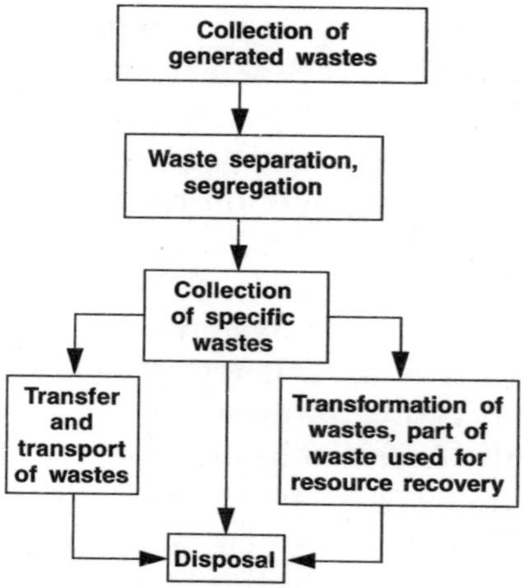

Fig. 5.2 Schematic representation of waste generation, collection and disposal

Waste production can be minimized by adopting the 3 R's: Reduce, Reuse, Recycle

Reduce the amount and toxicity of garbage and trash that you discard.

Reuse containers and try to repair things that are broken.

Recycle products wherever possible, which includes buying recycled products i.e. recycled paper books, paper bags etc.

These are processes that involve integrated waste management practices (IWM). They can reduce the wastes generated by approximately 50%.

1. **Reduction (Waste prevention)**

   Waste prevention, or "source reduction," means consuming and discarding less, is a successful method of reducing waste generation. Backyard composting, double sided copying of papers, purchasing durable, long-lasting environmentally friendly goods; products and packaging that are free of toxics, redesigning products to use less raw material in production and transport packaging reduction by industries are the normal practices used and have yielded substantial environmental benefits. Source reduction prevents emissions of many greenhouse gases, reduces pollutants, saves energy, conserves resources, and reduces the need for new landfills and combustors. It reduces the generation of waste and is generally preferred method of waste management that goes a long way toward saving the environment.

2. **Reuse**

   Reuse is the process, which involves reusing items by repairing them, donating them to charity and community groups, or selling them. Reusing products is an alternative to recycling because the item does not need to be reprocessed for its use again.

Some suggestions to reuse things

They include using durable glassware, steel ware, using cloth napkins or towels, reusing bottles, reusing boxes, purchasing refillable pens and pencils.

3. **Recycling**

   The process of recycling, including composting, has diverted several million tons of material away from disposal. Recycled materials include batteries, recycled at a rate of 93%, paper and paperboard at 48%, and yard trimmings at 56%. These materials and others may be recycled through drop-off centres, buyback programmes, and deposit systems. Recycling prevents the emission of many greenhouse gases that affect global climate, water pollutants, saves energy, supplies valuable raw materials to industry, creates jobs, stimulates the development of greener technologies, conserves resources for our children's future, and reduces the need for new landfills and combustors. For example, by recycling of solid waste in 1996, the United States prevented the release of 33 million tons of carbon into the air roughly the amount emitted annually by 25 million cars. Recycling can create valuable resources and it generates a host of environmental, financial, and social benefits. Materials like glass, metal, plastics, and paper are collected, separated and sent to processing centres where they are processed into new products.

Advantages of recycling

The advantages of recycling are it conserves resources for future generation, prevents emissions of greenhouse gases and pollutants, saves energy, supplies valuable raw, materials to industries, stimulates the development of greener technologies, reduces the need for new landfills and incinerators.

Disposal techniques

After sorting out the wastes for reuse and recycling, the materials for disposal are to be disposed off in a proper manner and in an eco-friendly way. Waste disposal is increasingly becoming a serious problem globally. Well-managed medical establishments have facilities for incineration of wastes. Majority of wastes land up on the roads or are piled up in open areas. Let us now have an overview of waste disposal mechanisms and technologies that are practiced.

1. **Open dumping**

This involves collecting solid wastes from different localities of a town and dumping them in low/lying areas generally on the outskirts of the town. These are uncovered pits and attract flies, insects, mosquitoes, bees, and rodents and emit obnoxious odours. This is not a proper waste disposal mechanism but practiced in many towns in India.

2. **Solid waste landfills**

Modern landfills are well-engineered facilities that are located, operated, designed and monitored to ensure compliance with local legal regulations. Pollutants from the sanitary landfills include volatiles like ammonia, methane, hydrogen sulphide liberated from anaerobic processes. The pollutants from leachates include heavy metals, organics, sulphates, nitrates etc. Therefore, solid waste landfills must be designed to protect the environment from contaminants that may be present in the solid waste stream.

There are different types of solid waste landfills:

Municipal solid waste landfills or sanitary landfills

Domestic residential, industrial wastes are diverted to solid waste or sanitary landfills. These are engineered structures where waste is spread out in 0.9 - 4.5 m thick layers and placed into the land, compacted and covered with clay (Fig. 5.3). Anaerobic bacteria execute the bio-degradation of wastes and methane is generated as by product. Landfills usually have liner systems and other safeguard mechanisms to prevent leakage and subsequent groundwater contamination. A few basic conditions should be met by any site, design and operation before it can be regarded as a sanitary landfill:

- **Location and design**: It is to be ensured that landfills are built in acceptable geological areas far away from faults, wetlands, flood plains, or other restricted areas.

Local geological and hydro geological investigations should be carried out before they are developed. The site should have additional lining materials to reduce leakage from the base of the site (leachate) and help reduce contamination of groundwater and surrounding soil. Leachate collection and treatment should be carried out carefully without polluting the surrounding environment. A waste disposal plan and a final restoration plan should also be included.

- **Liners:** This should include a flexible membrane overlaying two feet of compacted clay soil lining the bottom and sides of the landfill to protect the groundwater and underlying soil from leachate releases.
- **Leachate collection and removal:** This should be done effectively from the landfill for treatment and disposal.
- **Operating practices and cleaning:** They include compacting and covering waste frequently with several inches of soil help reduce odour; control litter, insects, and rodents; and protect public health. Control and clean up landfill releases and achieves groundwater protection standards. Trained technical staff should be recruited at the landfill to supervise site preparation, construction, depositing of waste and regular operation and maintenance.
- **Groundwater monitoring and post-closure requirements**: It requires testing groundwater wells to check if any waste materials have escaped from the landfill. Closure requirements include covering landfills and providing long-term care of closed landfills.

Advantages and disadvantages of municipal solid waste landfills

The natural resources are returned to the soil. As the soil becomes compost and fertile the wasteland is naturally converted to a useful land. Since they are covered, there is no problem of bad odours or insect breeding on them. A large area of land is required. Due to the production of methane, there may be fire hazards.

- **Bioreactor landfills**

  Municipal solid waste landfills that are designed to quickly transform and degrade organic wastes are known as Bioreactor Landfills. The increase in the rate of waste degradation and stabilization is accomplished through the addition of liquid and, in some cases, air to enhance the involved microbial processes.

- **Construction and demolition debris landfills**

  These landfills are characterized by input materials such as concrete, asphalt, brick, wood, asphalt roofing shingles, metals, and some types of plastics generated during the construction and demolition of homes, commercial buildings, and other structures. Due to the relatively inert nature of the debris materials, these landfills are subject to less stringent standards compared to municipal solid waste landfills.

- **Industrial waste landfills**

  These landfills are designed for the management of non-hazardous industrial waste that results from the production of various material goods and products.

3. Composting

Controlled biological decomposition of organic matter, such as food and yard wastes, into humus, a soil-like material is also known as composting. Composting is a bio-chemical process in which some microorganisms like fungi and bacteria, convert degradable organic waste into humus like substance. This end product, which looks like soil, is high in carbon and nitrogen and is an excellent medium for growing plants. It is the natural mechanism of recycling organic waste into new soil, which can be used in gardens, landscaping, and many other applications. There are two methods by which compost can be produced, aerobic (involving oxygen) and anaerobic (absence of oxygen) methods. The anaerobic process takes 4 – 12 months for decomposing and it produces odours. The aerobic method is the best preferred and compost is rapidly produced. The following steps are used in composting process.

- **Sorting** - Substances like tins, glass, plastics, cloths etc are manually removed.
- **Shredding** – The waste is chopped or shredded into smaller pieces in order accelerate the rate of decomposition and aeration. The pieces range from 25 to 50 mm.
- **Windrow system** – The shredded waste is stacked in rows of 1.5 – 1.8 m height and approximately 3 m width. They are kept in the open for bacterial activity to occur. The temperature increases and the wastes start decomposing within 10 – 30 days and the colour of the waste also changes to a brown peaty material.
- **Turning** – The windrow mounds are frequently turned for good aeration, normally once in 3 days.
- **Bottom aeration** – Here air is forced up from the bottom to facilitate aerobic decomposition.

The MSW generated in India comprises of 35%–40% of organic matter. Apart from being clean, cheap, and safe, composting can significantly reduce the amount of disposable garbage. The organic fertilizer can be used instead of chemical fertilizers and is better specially when used for vegetables. It increases the soil's ability to hold water and makes the soil easier to cultivate. It helped the soil retain more of the plant nutrients. In vermi-composting, earthworms are added to the compost. They help in degrading the waste and the excreta of the worms make the compost very rich in nutrients.

Mechanized composting

Compost can also be produced using equipments. It is much more efficient and it facilitates easy handling of the waste. Once trucks collect the waste, they are processed as follows,

- **Picking belt** - which removes the plastics, glass, cloth etc.
- **Pulverizers** – that shreds the waste
- **Vibrating screens** – that removes large materials from the waste
- **Ballistic separators** – that remove dense particles
- **Windrow composting machine** – that mixes and piles the compost in triangular shaped rows
- **Rotating agitators** – mix the compost by turning and tossing it
- **Dryers** – that dry the composted material
- **Packaging** – the compost is mixed with nutrients and packaged.

Mechanized composting plants are found in Mumbai, Bangalore, Ahmedabad, Kolkata and Delhi.

Advantages of composting

Keeps organic wastes out of landfills, provides nutrients to the soil, it makes soil easier to cultivate, it helps reduce the adverse effects of excessive alkalinity, acidity, or the excessive use of chemical fertilizer, increases beneficial soil organisms, i.e. earthworms and centipedes, reduces the need for fertilizers and pesticides, protects soils from erosion, it helps in controlling the growth of weeds in the garden, assists in pollution remediation.

4. Incineration

Burning of solid wastes at very high temperatures (around 1800°C) is called incineration. An incinerator is a high-temperature furnace used for burning refuse.

The old incinerator models were simple brick-lined enclosures as compared to the modern rotary-kiln incinerators which are used by modern industries. They have a long, inclined passageway through which refuse is continuously moved. They can generate energy while reducing the amount of waste (approximately up to 90 percent in volume and 75 percent in weight). Only 10% of ash remains in the process.

Switzerland, Sweden, Germany, Japan, Netherlands, Italy and Luxembourg incinerate more than 50% of their municipal waste. They have their incinerators equipped along with a flue gas treatment facility.

The main components are: the rotary kiln, afterburners and air pollution control systems. The rotary kiln is the main combustion chamber for both solid and liquid wastes. Afterburners are secondary combustion chambers which destroy materials that were not incinerated during the initial process. The products of combustion are carbondioxide, nitrogen, water vapour and ash. Afterburners help in breaking the chemical bonds of the waste compounds. All the processes are designed, controlled and monitored to assure safe operations. The inorganic ash residue from the rotary kiln is collected in a large container. This ash and the residue from the air pollution control systems, are then tested for hazardous material concentration levels. Subsequently, they are mixed with common chemical stabilizers in order to conform to the levels specified by the Environmental Protection Agency for safe landfill disposal.

Types of incinerators

Incinerators are typically classified on the type of fuel they burn in their operation. **Mass burn incinerators** do not segregate the waste to be destroyed. Other types of incinerators destroy waste produced from recycling processes. **Hazardous waste incinerators** destroy medical waste and other types of waste which is potentially contagious or harmful for the environment.

Advantages and disadvantages

One of the main advantages is that reduction in the quantity of waste products takes place prior to their final disposal. While incinerators can operate without being used for energy production, the steam heat generated by incinerators produce as much as 100 MW electricity and is being used by some municipalities to utilize this as an electrical source. This allows for the recovery of energy for electrical and heating purposes and reducing the use of fossil fuel energy sources. Incineration is most effective in destroying specialized wastes like hospital and biological refuse. It is the safest method in terms of hygiene. The main disadvantage is that incineration releases toxins and dioxins into the atmosphere causing serious concern. Two types of residues are produced which is hazardous: fly ash and bottom ash.

Other new technologies for waste disposal (from World Resources Institute, 1992)

- **Plasma arc destruction** – It involves creating very high temperatures (6000 - 15000°C) and a plasma state. This is successful in destroying PCBs, organic pollutants and other hazardous substances. The disadvantage is that it is very expensive.
- **Biotechnology** – It involves the use of specifically cultured, genetically engineered or using appropriate microorganisms in the composting process. It is a very slow process and is not successful in destroying all the hazardous compounds.
- **Molten salt destruction** – It involves treating the toxic wastes like DDT, acids, corrosive agents etc. with molten salts at high temperatures (1650°C). Hexachlorobenzene can be destroyed to 99.999% by treating it with a molten mixture of Na_2CO_3 and Na_2SO_4 at 900°C.
- **Superheated water** – This involves dissolving organic compounds in aerated water that is heated to 370°C

at 220 Bars pressure. The organic pollutants are oxidized to CO_2 and H_2O.

Recovery and conversion methods

Converting trash to useful resources

Waste can be converted to useful resources if treated properly. Some examples are given below wherein different types of waste or trash are transformed into a resource. The concerns about petroleum reserves and the environmental consequences (for eg. global warming, climate change) of fossil fuels and have urged man to search for alternate/ renewable transportation bio fuels. Typically, a bio fuel should provide a net energy gain, have environmental benefits, be economically competitive, and be producible in large quantities without reducing food supplies. The important bio fuels experimented include: ethanol from corn grain, bio diesel from soybeans, biogas from manure, cow dung, landfill gas from solid wastes, hydrogen gas etc. Compared to the fossil fuels they reduce the greenhouse gas emissions (ethanol by 12% and bio-diesel by 41%).

1. **Compost :** Microorganisms aid in the biological degradation of organic waste and it results in an organically rich soil like product, called compost. This is mixed with nutrients and packaged for marketing. It can be used in gardens and agricultural fields for enriching the soil.
2. **The anaerobic nutrient rich waste waters :** The anaerobic waste water treatment processes yield nutrient rich waste waters which could be used to cultivate ornamental plants, for example roses. By this process we can produce ornamental, decorative plants of commercial value.
3. **Vermicomposting :** Organic materials constitute two-thirds of the total solid wastes, out of which a large portion is food waste. The vermicomposting process involves using earthworms (*Eisenia foetida)* at

temperatures between 50° and 70° F to convert organic wastes into compost. "Vermi" means worm in Latin. The worms like to feed on decomposing organic materials (e.g., kitchen wastes, litter, etc.) When done properly, the worms transform the food wastes into nutrient-rich castings. The product, called castings, is highly rich in beneficial microorganisms and nutrients. This makes it an organic safe fertilizer. This is a good way to convert trash into resource.

4. **Landfill gas-an alternate source of energy :** Landfills are one of the largest sources of methane emissions. Methane is produced in an anaerobic environment by the bacterial decomposition of organic materials such as yard waste, household waste, food waste, paper etc. For example, in 1996, 14 % of the methane generated in U.S. landfills was recovered and combusted for energy. The U.S. Environmental Protection Agency is currently reviewing site-specific information on landfill gas recovery and anticipates that this will lead to greater methane recovery nationwide and lower the net methane emissions. Energy recovery can capitalize on the energy value of landfill gas and replace the use of fossil fuels. Large scale and small-scale industries, research institutions, hospitals, and other large energy consumers can greatly benefit by tapping landfill gas-a local, renewable energy resource. The use of landfill gas for energy offers significant environmental, economic, and energy benefits to landfill owners and operators, project developers, energy product purchasers and consumers, and communities living around landfills.

Advantages of landfill gas to energy projects

It would create jobs, it would help to reduce explosion hazard at landfills, it would help in reducing emissions of methane, a powerful greenhouse gas, formation of smog will be reduced, displace electricity produced by fossil fuels, further reducing pollution.

Resource Conservation and Recovery Act

The Resource Conservation and Recovery Act (RCRA) was enacted in 1976 and amended in 1984 in the U.S. RCRA calls for conservation of energy and natural resources, reduction in the waste generated, and environmentally sound waste management practices. The primary objective of the act is to protect human health and the environment from the potential hazards of waste disposal. Resource recovery means that treatment processes would yield resources, for example methane gas, compost, etc. which is eco-friendly.

Case Studies: Reuse of Solid Wastes

1. **Reuse of chrome bearing solid waste in Chennai**

   Chromium metal is one of the important toxic heavy metals in tannery wastes. Central Leather Research Institute, Chennai have come up with a new idea to remove chromium from tannery effluents and to reuse the chrome bearing solid waste. The technique involves using, a common seaweed, *Sargassum wightii* which when added to the tannery effluent removed chromium within six hours. The reported maximum uptake was 35 mg of chromium per gram of seaweed. The chrome bearing solid waste was used to make chromium sulphate, a major tanning agent thus reusing the waste product.

2. **Wealth from furniture Waste in Sri Lanka**

   Moratuwa, a famous furniture-manufacturing centre in Colombo, Sri Lanka is dumped with tons of furniture waste materials. Everyday 20 tons of wood waste is dumped here and also into the nearby oceans. A group of furniture manufacturers are reusing the wood waste to make chipboards. 15 tons of waste wood including wood chips, particles and wood dust is recycled per day and 400 chipboards are made from this thus saving precious forest resources.

Measures to reduce solid waste

- Carry a cloth or jute bag when you go to the market.
- Make a compost pit in your garden and put the biodegradable wastes into it.
- Segregate your waste at home – keep separate garbage bins and dispose organic, plastic containers, polythene, plastic bags, glass, textile and paper separately. This is being practiced by the developed nations like Switzerland, Germany, U.K., U.S.A and majority of the European nations.
- When you go out do not throw paper and other wrappings on the streets. Dispose them in a dustbin.
- Sell all the recyclable items that are not of use to a person who trades in scrap items.

Some Interesting Facts

- Millions of trees and bamboo plants are being cut every year in China to produce 45 billion disposable wooden chopsticks, which are exported as well as used domestically. With this current rate of timber use, China would have no forests left very soon.
- Severe soil erosion because of poor land use management and lack of geological understanding has lead to a decline in agricultural productivity and desertification in Spain. It is the most arid part of Europe.
- A baby's diaper takes 500 years to decompose, as it is non-degradable. In the US alone 18 million diapers end up in landfills every year. One billion trees per year are used worldwide to make these diapers.
- Polythene covers and bags cause harm to the environment by blocking the drains and animal intestines. An enormous number of cattle die in our capital, New Delhi due to their consumption of plastic bags thrown in the dustbins.
- 4000 plastic bags were recovered in a stomach of a cow in Bhuj by a veterinary surgeon, Dr Jadeja, who is a founder member of the Society for the Prevention of Cruelty to Animals.

- Himachal Pradesh is the first state in India to have banned the use of polythene, plastic covers. Here, the shopkeepers and the people use only paper covers, paper bags and jute bags.
- A swamp deer and a deer found in Manipur also died due to the consumption of plastic bags at the Delhi zoo.
- India's first ever power plant using solid waste as fuel is commissioned at Perungudi in Chennai, Tamil Nadu. The project is being undertaken by Australia-based Energy Development Ltd and is of the 1.5-billion-rupee 15 MW project.
- The European Union has banned the use of six phthalates in toy manufacture as children can absorb them by chewing or biting toys containing the same. It causes cancer, asthma and allergies.
- Mr. Nek Chand has created an exquisite rock garden at Chandigarh, which is entirely made from waste. Each item in this garden has been beautifully crafted from wastes like – bottles, tin, tyres, plastics, tube lights, etc. This exquisite rock garden is a popular tourist spot in the city.
- 4 - nonylphenol, a chemical used in cleaning materials, textiles, plastics etc. acts as a trigger for breast cancer.
- Cotton stalks, straw and other bio wastes can be transformed into compost by farmers.
- Scientists from U.K. suggested that recycled paper compost could help in fight of plant diseases. When 20% of the soil is replaced by paper compost, the incidence of diseases is drastically reduced. For example the tomato root-rot and wilt diseases was reduced by 72% and 61% respectively.
- More than 2300 tons of construction debris is generated everyday in Mumbai. This is a serious problem of solid waste in the city. The Youth for Unity and Voluntary Action (YUVA) with the support of City and Industrial Development Corporation (CIDCO) have recycled the waste into construction material, bricks and interlocking pavers.

UNIT 6

NOISE POLLUTION

The word noise comes from the Latin word nausea. Unlike water, air and soil pollution, noise pollution does not leave any residue. Noise pollution is any undesirable or unpleasant sound that causes discomfort to the ear and results in consequent health hazards. Noise pollution is more prevalent in cities compared to the suburbs and villages. In larger cities the reason behind noise is the increasing population and lifestyle impacts. More people in a city mean more public transportation, more industries, institutions, factories, and of course more residential areas that collectively produce large amount of noise .The increase in environmental noise is a by-product of technological advances.

The WHO (World Health Organization), in a report to the UN environment conference, Stockholm (1972) stated "Man is fast driving himself mad with noise. Of all forms of pollution, noise is the most insistent yet the easiest to control". The report described noise as "a curse of modern times", and "a major environmental problem today".

Switzerland is one of the countries that have incorporated the best possible technology for reducing noise. This beautiful alpine country is almost noise free. This is because of the careful planning of the city and the co-operation of the citizens. The Swiss society of engineers and architects has defined the required norms for soundproofing building interiors: stairwells, lifts, heating and ventilation systems.

NOISE : CASE STUDY

Noise hits whales in Hong Kong

Studies have shown that shipping traffic in Hong Kong, which is one of the busiest ports in the world with approximately half a million oceanic vessels travelling through its waters every year (including over 10,000 transits by high speed ferries) has caused changes in the dolphin and whale behaviour especially in response to fast moving vessels. A special sanctuary was established by the Hong Kong government in 1995, surrounding the islands of Sha Chau and Lung Kwu Chau, an important place occupied by the *humpback dolphins*. At any given time approximately 200 vessels surround this sanctuary. The sanctuary was a measure to mitigate boat traffic and tremendous noise produced. Adjacent to the sanctuary is an airport, where 700 planes descend and take off everyday, directly over the sanctuary. All the above activities have caused high noise input into the natural whale habitat. Noise, a major anthropogenic stress factor has caused a general decline in the whale populations.

Human ear and sound

The visible outer part of the ears called the pinna gathers sound. When sound reaches the eardrum, they become vibrations. These vibrations are picked up by three tiny linked ear bones in the middle ear and are mechanically passed on to the inner ear where hair cells translate them into electrochemical impulses (Fig. 6.1). The auditory part of our brain interprets this stimulus or the electrochemical impulses and tells us what to do.

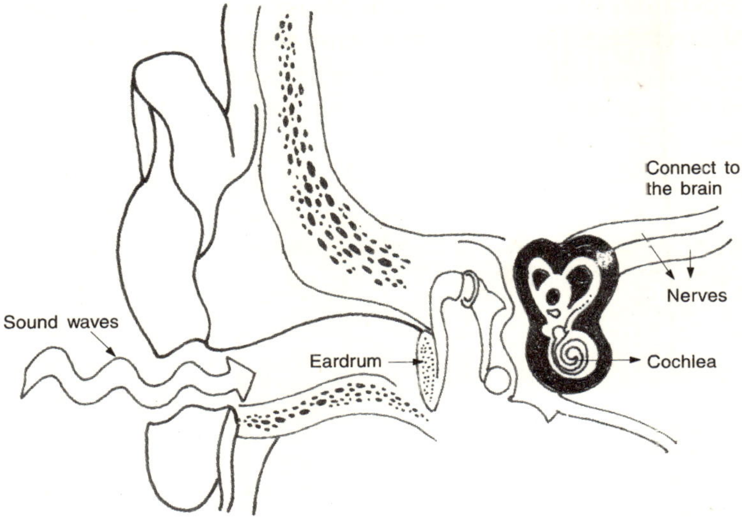

Fig. 6.1 : Anatomy of human ear

Difference between sound and noise

What may be a disturbing noise for one person may be a pleasant sound for another person. Sound is a compression waveform that travels through matter. It can readily travel in water, steel, air but it does not travel through vacuum. Sound waves are characterized by amplitude, frequency, velocity and wavelength. When sound hits an object the object vibrates. Sounds that are generally considered to be pleasant create variations in pressure with a regular pattern. Sound waves that are noisy or called as noise have irregular patterns. The patterns created by human voices have both irregular and regular sound pulses. Noise is produced by fluctuations in air pressure, and includes sound and vibration. It is not the pitch or loudness of a sound that makes it unbearable but it is the repetitive nature, the distraction it causes and the lack of control over it that makes sound a noise. The characteristics of unpleasant sound depend on the frequency of sound waves, the intensity of sound waves, time and exposure of sound waves and the intermittence of sound waves.

Noise is characterized by loudness measured by decibel (dB) and pitch or frequency measured in hertz (Hz). An adult with good hearing can hear frequencies between 20 and 20,000 hertz (Hz); frequencies between 2000 Hz and 8000 Hz is considered most annoying. Those below 500 Hz are considered less annoying. The minimum audible sound is 10 dB, a whisper is 20 dB and a normal conversation is between 35 – 55 dB. A jet plane takes off at 150 dB and a rocket engine at 180 dB. The World Health Organization (WHO) has fixed the safe level for noise in cities as 45 dB.

Sources of noise pollution

Each day noise pollution affects the health of millions of people. Majority of people are exposed to high levels of noise. It could be from loud music, a motorcycle, aeroplane, or the work place. Transportation vehicles produce the maximum amount of noise in our environment. It is reported that continuous exposure to noise with a volume more than 70 decibels can lead to permanent hearing loss. Constant noise can affect a person physically and mentally. This has become a significant public health issue. The sources include: factories, industries, trains, buses, automobiles, aeroplanes, radios, televisions, loudspeakers, etc. Barking dogs, lawn mowers, electrical drillers, power saws, snow blowers, church bells, jackhammers, motorcycles, airplanes, stereo systems, and traffic generally induce irritation, annoyance, discomfort, and hearing impairment.

Table 6.1 : Some sources of sound and their sound level

Sources	Sound Level in dB
Rockets	180 dB
Jet plane take off	140 dB
Aeroplane take off,	120 dB
Industrial grinders	90 – 115 dB
Racing cars	110 dB

Contd...

Contd...

Sources	Sound Level in dB
Garbage Trucks	100 dB
Motor cycle	90 dB
City Traffic	60 – 90 dB
Hydraulic pumps	80 dB
Vacuum cleaners	70 dB
Talking (normal)	40 – 50 dB
Whispering	30 dB
Flowing streams	15 dB
Rustling of a leaf	10 dB

Noise can be divided into industrial, transportation and domestic noise.

- Industrial noise: factories, businesses, extractive industries, construction, agricultural and other industries
- Transportation: air, road, rail and shipping activities
- Domestic noise: using noisy articles, vehicles and vessels, barking dogs and loud music.

Industrial

Industries create lot of sound, for example even automatic filling of oil cans in the oil refinery can cause considerable sound. Some of the industrial sources include metallurgical operations, motors, agitators, pneumatic power tools (grinders, chippers), moulding machinery, combustion (furnaces, flare stacks), air compressors, hydraulic pumps, turbo-generators, blowers, industrial trucks, transformers, textile mills, printing presses, etc.

Transportation

- **Road traffic**

  Automobiles and road traffic noise is one of the largest source and a widespread environmental problem. Motor cycles, buses, trucks, tractors, lorries are increasingly becoming an important source of high intensity noise. The impact of road traffic noise on the community depends on various factors such as road location and design, land use planning measures, building design, vehicle standards.

- **Air traffic**

  The extent of air traffic and aircraft noise impact depends on the types of aircraft flown, the number of flights, location of the aerodromes, design and land use planning patterns and flight paths.

- **Rail traffic**

  Rail traffic is another cause of noise pollution in India. Indian railways run more than 1000 express and mail trains per day connecting major cities/ states. Most of the trains are very old and run on diesel. The level of noise associated with rail traffic is related to the type of engine or rolling stock used the speed of the train and track type and condition. Electric trains are generally quieter than diesel trains. Areas affected by freight trains often experience higher noise levels than areas affected by passenger trains. There are two main sources of noise and vibration relating to the operation of the rail network: the operation of trains and the maintenance and construction of rail infrastructure.

- **Domestic noise**

  Other significant sources of noise annoyance include barking dogs, car alarms, garbage recycling, lawn-mowers, building construction, drilling operations, grinders, mixies, televisions, radio, music systems, cookers, plumbing and household noise. The national noise survey, USA found that noise from barking dogs and road traffic have the greatest impact on residential communities. Noise from domestic animals (dogs, donkeys, cows, hens etc.) is of particular concern because it is unpredictable and often occurs repeatedly.

- **Other significant sources**

  Satellites are launched into space with high explosive rockets. This causes a deafening noise especially at the time of lifting off, of the satellite. Again tons of trinitrotoloune (TNT) and explosives are used that cause a lot of noise pollution. Defence, artillery equipments, military airplanes are other sources that contribute to noise pollution.

Effects of noise pollution

The effects of noise pollution on human beings can be broadly classified into two categories: i) Auditory effects and ii) Non-auditory effects

Auditory effects

- Neural hearing loss is caused by damage to auditory nerves; sensory hearing loss is caused by damage to the inner ear and is most commonly associated with noise.
- High noise levels can result in excessive ear wax secretion, a damaged ear drum or fluid in the middle ear.
- More serious effects include permanent deafness, pain and ringing in the ears

Non-auditory effects

- A high noise level can also lead to increase in blood pressure and blood cholesterol, increase in heart rate, dilation of pupil of the eye, adrenalin hormone changes, digestive spasms, damage to the respiratory and digestive systems.
- Constant noise can lead to stress disorders, which could develop into ulcers.
- It can contribute to mental illnesses, depression, hypochondria and sexual dysfunction.

- Sleep deprivation and sleep disorders occur at 45 dB or higher and at 85 decibels hearing damage is most likely to occur.
- It can cause irritation, heart attacks, constriction of blood vessels, insomnia, anxiety, sweating, giddiness, nausea, gastrointestinal problems, peptic ulcers, etc.
- In children it can cause a decrease in alertness and the ability to memorize, nervousness, pupil dilation and a decrease in the visual field.

Effects on wildlife

Noise can have adverse effects on wildlife. The Delhi zoo authorities noticed physiological and psychological effects on the deers, lions and rhinos. A decline in the migratory birds was also observed.

Noise control measures

The general public must be made aware of the problem and seriousness of noise pollution. This could be achieved through radio, television, newspapers, or community gatherings. Teachers and parents can teach children to lower their televisions and stereos in their rooms thus further preventing noise pollution. Society and people have the ability to make the greatest changes concerning noise pollution. Noise reduction or mitigation measures are essential. Some of the control measures are adopted are to control noise levels generated from a source (vehicle/ machine design), to control noise transmission (noise barriers), to reduce the level of sound reaching a receiver (soundproof buildings).

- **Reducing domestic noise** – The residences can also be made sound proof by using heavy curtains on the windows, putting rugs / carpets on the floors, and also using acoustical tiles on the ceilings and walls. Again, planting trees and shrubs in front of homes and buildings absorb sounds.

- **Reducing road traffic noise** – Old vehicles may be replaced, less use of sirens and horns, proper oiling of engines, making new or upgraded roads, construction of one-ways, bridges or flyovers for regularizing the traffic, proper functioning of the traffic lights to avoid honking of horns. Noise pollution, especially concerning transportation, can be reduced by banning car horns in certain areas and by limiting the traffic flow through residential areas. Vegetation buffer zones must be created in different parts of the city. Efforts should also be made for roadside plantations. Noise absorbing trees like ashoka, banyan, neem, casuarinas, coconut and tamarind should be used for roadside plantations.
- **To reduce noise in sensitive locations** such as residences and schools. These include using noise mounds, noise attenuation walls and quieter road surfacing.
- **Reducing air traffic noise** – design and construction is important, aerodromes should be away from residences, schools, hospitals, construction of smooth runways.
- **Reducing railway noise** – retrofitting existing locomotives to reduce noise emitted, upgrading existing track to continuously welded rail which removes rail joints which is a significant source of noise and vibration, designing new bridges to reduce noise and retrofitting of existing bridges with noise attenuation devises, deploying quieter rolling stock in noise sensitive areas, use of electric locomotives, altering the holding pattern of trains to avoid them being held at signals for extended periods.
- **Reducing industrial noise** – Noise pollution can be controlled in industries by using sound absorbers, acoustic tiles, silencers in engines and home appliances. Workers at factories should wear earmuffs or earplugs to decrease the amount of noise levels.

- **Reducing noise in schools** - building finishes and construction details designed to minimize noise impacts in sensitive areas such as music rooms, technology workshops and gymnasiums, a range of solutions to deal with road traffic noise, including acoustically sealing walls closest to the noise source, mechanically ventilating rooms and construction of barriers between the noise source and the affected site, measures to deal with air traffic noise, including orientation of rooms to reduce noise exposure; use of noise attenuating materials and construction techniques; larger than normal roof overhangs and heavily insulated roofs with acoustically absorbent linings; and soft finishes to the ground immediately adjacent to windows to reduce sound reflection into openings.
- **Making a green** belt preferably around aerodromes, train stations, vegetation cover as plants absorb noise and can deflect noise upward.
- **Individually** by using ear plugs, and creating awareness among the public.

Noise measuring instruments

Noise measurement is a first step for applying noise control technology.

- **Sound level meter** is an instrument, which indicates sound pressure levels in the audible range. A sound level meter consists basically of a microphone and an electronic circuit including an attenuator, amplifier, weighting networks or filters and an indicator meter or display unit.
- **Impulse meter** Noises lasting a fraction of a second such as those produced by hammer blows are measured with impulse meter, as the sound level meter is too sluggish to indicate peak levels of transient noise.

- **Frequency analyzers** are also used, which allows separating the main components of the signal by dividing the frequency range of interests into smaller frequency bands using a set of filters.
- **Graphic recorders** are also used where a graphic record of a sound level is obtained by connecting a sound level meter to a graphic level recorder that plots the sound pressure level as a function of time on a moving chart paper.
- **Noise dosimeters** not only tell us about the sound levels but also the amount of time a worker is exposed at various levels of noise.

Noise control technologies

There are various technologies that can be incorporated for noise reduction at source. Some of the technologies are given below.

- **Damping** is a method of noise reduction at source that reduces noise radiated from structures vibrating in resonance by employing vibration absorbers, monolayer absorbing linings.
- **Sound absorbers or silencers** are noise-reducing devices that diminish air borne sound propagation in ducts or which prevent sound generation in vents or openings when gases escape or expand.
- **Acoustical enclosures** are full or partial enclosures when provided to noisy machines resulting in the reduction of noise. The enclosure may be made of sheet metal lined inside with an acoustical material.
- **Acoustical barriers or shields** The erection of acoustical barriers or shields between the noise source and other work places is effective in controlling the direct radiation of noise as high frequency noise behaves like light beam or radiant heat as far as reflection is concerned.

5. **Acoustical treatment of walls and ceilings** Sound energy is absorbed whenever it contacts a porous material and usually absorbed 50 – 90 % of the incident sound energy depending on its frequency and angle of incidence. Acoustical materials may be absorbent fibres, glass or mineral wool and some hard but porous plaster like materials.

Standards

The following steps were taken by the Indian Government in order to control noise pollution.

- Ambient noise standards were notified in 1989, which formed the basis for State Pollution Control Boards to initiate action against violating sources.
- The vehicular noise standards, notified in 1990, are being implemented by the Ministry of Science and Technology, to reduce traffic noise. These standards have been made more stringent vide a notification in September 2000, effective from January 2003.
- Noise standards for diesel gensets were prescribed in December 1998. Govt. has been pursuing with State Pollution Control Boards, generator manufacturing and major users, for implementation of these standards. Noise standards for petrol and kerosene generator sets were notified in September 2000, and will be effective from September 2002. The sale of these gensets will be prohibited if not certified by the testing agencies, identified for the purpose.
- Noise standards for fire-crackers were developed in October 1999. Central Pollution Control Board had carried out a compliance testing of the fire crackers available in the market and also taken up with the Department of Explosives for compliance with these standards.
- The Noise Rules, 2000, regulates noise due to Public Address System/ Loud speakers and also prescribed procedures for noise complaint handling.

Table 6.2 : Ambient air quality standards in respect of noise (CPCB)			
Area code	**Category of zone**	**Noise Limits in dB(A) Leq ***	
		Day time	Night time
(A)	Industrial area	75	70
(B)	Commercial area	65	55
(C)	Residential area	55	45
(D)	Silence Zone	50	40

Some of the important points of the act are given below (CPCB)

- Day time shall mean from 6.00 a.m. to 10.00 p.m.
- Night time shall mean from 10.00 p.m. to 6.00 a.m.
- Silence zone is defined as an area comprising not less than 100 metres around hospitals, educational institutions and courts. The silence zones are zones, which are declared as such by the competent authority.
- Mixed categories of areas may be declared as one of the four above-mentioned categories by the competent authority.
- *dB(A) Leq denotes the time weighted average of the level of sound in decibels on scale A which is related to human hearing.
- A "decibel" is a unit in which noise is measured.
- "A", in dB (A) Leq, denotes the frequency weighting in the measurement of noise and corresponds to frequency response characteristics of the human ear.
- Leq: It is energy mean of the noise level, over a specified period.
- A loud speaker or a public address system shall not be used except after obtaining written permission from the authority. A loud speaker or a public address system shall not be used at night (between 10.00 p.m. to 6.00 a.m.) except in closed premises for communication within, e.g. auditoria, conference rooms, community halls and banquet halls.

Some Interesting Facts

- Switzerland and Australia are the most advanced countries with noise reduction regulations.
- The Indian government has banned 20 types of crackers that exceed 125 decibels and any person who manufactures and stores them will face prosecution.
- About 80 per cent of the fire-works sold in New Delhi come from Sivakasi in Tamil Nadu while the rest comes from Uttar Pradesh.
- In Australia, even lawn mowers, chain saws and jackhammers must carry a "noise" label.
- Noisy cars are suspended off their registration until they are repaired in an authorized garage in Adelaide, a city in southern Australia.
- In New South Wales and in some parts of the U.S., noisy vehicles can be stopped and tested by the side of the road.
- Studies show that the radiation from cell phones affect the brain, ears, eyes and disturb sleep.

References

Agency for Toxic Substances and Disease Registry (ATSDR). 2000. Toxicological Profile for polychlorinated biphenyls (PCBs). Atlanta, GA: U.S. Department of Health and Human Services, Public Health Service.

Agrawal. K.C. 1999. Environmental Biology, Agro-Botanica.

Alloway B.J and Ayers D.C. 1993. Chemical principles of environmental pollution. Blackie academic and professional, an imprint of Chapman and Hall.

Baljeet S. Kapoor. 1997. Environmental Engineering An overview. Khanna Publishers

Burks, B.D. and M.M. Minnis. 1994. Onsite Wastewater Treatment Systems. Madison, WI: Hogarth House, LTD.

Cesare Emiliani 1992. Planet Earth - Cosmology, geology and the evolution of life and the environment. Cambridge University Press U.K.

Down to Earth Issues 2004, 2005 and 2006.

Edward A. Keller 2000. Environmental geology, Prentice Hall, Inc.

Hoornweg, Daniel with Laura Thomas. 1999. Working Paper Series Nr. 1. Urban Development Sector Unit. East Asia and Pacific Region.

Kamala. A and Kantha Roa D.L. 2001. Environmental engineering. Water supply, sanitary engineering and pollution. Tata McGraw-Hill publishing company limited.

Karen Mancl Wastewater Treatment Principles and Regulations AEX-768-96 Ohio State University Extension Fact Sheet, Food, Agricultural and Biological Engineering, 590 Woody Hayes Dr., Columbus, Ohio 43210

Landfill methane recovery, January 2000. Climate Change Technologies

Mark J. Hammer and Mark J. Hammer Jr. 1998. Water and wastewater technology Prentice Hall India.

Masters, G.M. Introduction to environmental engineering and science, Prentice Hall of India Pvt Ltd 1991.

National Small Flows Clearing house. 1997. Basic wastewater characteristics. Pipeline 8(4): 2-6

Nebel B.J. and Wright R.T. Environmental Science Prentice Hall Inc 1987.

Odum, E.P, Ecology. The Link between the natural and social sciences, IBH Publishing Company Delhi

Reflections on Water, National Geographic Society Teachers Handbook, Geography: Nov. '92.

Solid Waste Landfills: Decision-Makers Guide Summary. Thurgood, Maggie. 1999 World Bank, World Health Organization, Swiss Agency for Development and Cooperation, and Swiss Center for Development Cooperation in Technology and Management.

U.S. Environmental Protection Agency, Solid waste landfills

Wegelin, M., Boller, M., Schertenleib, R. (1987). Particle removal by horizontal-flow roughing filtration, J. Water Supply Research and Technology - Aqua, 36, 80-90.

Wegelin, M., Schertenleib, R., Boller, M. (1991). The decade of roughing filters - Development of a rural water treatment process for developing countries. J. Water Supply Research and Technology - Aqua, 40, 304-316.